OCEANOGRAPHY AND OCEAN ENGINEERING

SEAWATER

GEOCHEMISTRY, COMPOSITION AND ENVIRONMENTAL IMPACTS

OCEANOGRAPHY AND OCEAN ENGINEERING

Additional books in this series can be found on Nova's website under the Series tab.

Additional e-books in this series can be found on Nova's website under the e-book tab.

OCEANOGRAPHY AND OCEAN ENGINEERING

SEAWATER

GEOCHEMISTRY, COMPOSITION AND ENVIRONMENTAL IMPACTS

MATHEW R. WHITE
EDITOR

nova publishers
New York

For permission to use material from this book please contact us:
Telephone 631-231-7269; Fax 631-231-8175
Web Site: http://www.novapublishers.com

Library of Congress Cataloging-in-Publication Data

Seawater : geochemistry, composition and environmental impacts / Mathew R. White, editor.
pages cm
Includes index.
ISBN 978-1-62808-606-5 (hardcover)
1. Seawater--Composition. 2. Seawater--Analysis. 3. Chemical oceanography. 4. Marine bacteria. 5. Marine pollution. I. White, Mathew R.
GC111.2.S42 2013
551.46'6--dc23

2013028316

Published by Nova Science Publishers, Inc. † New York

CONTENTS

Preface		**vii**
Chapter 1	Impact of Internal Wave Breaking on the Ocean Upper Layer Formation *Alexander Yu. Benilov*	**1**
Chapter 2	Trace Elements in Seawater in *Offshore Oil Production Platform* Areas *Sergio Felipe Jerez Veguería and José Marcus Godoy*	**53**
Chapter 3	Self-Purification of Seawater: A Measure for Environmental Regulation *Sergei B. Gulin and Viktor N. Egorov*	**93**
Chapter 4	Macro-Imagineering a "Fifth Coast" USA: Salton Sea Megaport and Seawater Canal *Richard Brook Cathcart*	**127**
Chapter 5	Assessing the Effect of Nearshore Currents over the SO_2 Absorption in Caribbean Seawater *C. Riverol and T. Alvarez*	**157**
Index		**167**

PREFACE

In this book, the authors present current research in the study of the geochemistry, composition and environmental impacts of seawater. Topics discussed include the impact of internal wave breaking on the ocean upper layer formation; trace elements in seawater in offshore oil production platform areas; self-purification of seawater; the Salton Sea megaport and seawater canal; and assessing the effect of nearshore currents over the SO2 absorption in Caribbean seawater.

Chapter 1 – The physical models of air-sea interaction mainly concentrated on the processes at the air-sea interface, surface waves, mixing within the ocean upper layer and the formation its hydro-physical structure including its lower boundary. The last one is relatively narrow seawater layer with the sharp seawater density increase with depth growth. This layer is often called the pycnocline. The dynamical and thermo-dynamical contribution of the pycnocline into the upper layer is routinely ignored because the very strong density stratification makes the lower boundary impermeable for dynamical and thermo-dynamical interactions with the main ocean body below the pycnocline. The exception is only internal waves propagating at the pycnocline. However the internal waves, like the surface waves, break at randomly distributed locations at the pycnocline generating turbulent patches which violate the pycnocline integrity and so its impermeability. The paper describes the effect of internal wave breaking turbulence on the structure of the hydro-physical and biogenic fields near the ocean surface. The exchange of turbulent energy, heat and biogenic salts between the ocean upper layer and the main body of the ocean below is caused through the random patches of active turbulence within zones of internal wave breaking. The variability of the upper layer temperature, concentration of biogenic salts and the turbulent energy is caused by the turbulence of the breaking zones, its variability and the breaking zones statistics. The turbulent horizontal diffusion of the ocean upper layer spread the non-uniformities of these fields into larger horizontal scales than the breaking zones scales. Estimates of the manifestation of the discussed mechanism are shown in the mean values and variances of the surface non-uniformities of the upper layer fields as well in their spectrums. The simple model examples of the field of internal wave breaking turbulence allow quantitative evaluation.

Chapter 2 – Produced water is the largest wastewater discharge in offshore oil, gas exploration and production processes. This water consists of formation water, the water naturally present in the reservoir, and floodwater injected into the formation. In general, in offshore operations, seawater is injected into the reservoir to maintain pressure in the formation for oil transportation. The produced water, that usually has a higher salt concentration than seawater, is separated from the oil and then discharged to the surface water. In this chapter, the characterization and evaluation of the trace elements and natural radionuclides in seawater in the *Bacia de Campos*, or Campos Basin, oil field region, in Brazil, is presented. In the present work, a representative region of oil-bearing activity in the *Bacia de Campos*, an oceanic region of the oil-field offshore platforms, represented by two fixed production platforms, was studied. The element and natural radionuclide content in dissolved and particulate produced water and seawater around the effluent discharge sites were studied. Six samples were collected at three different depths (surface (5 m), middle (45 m) and bottom (95 m)) in the oceanic region, with the sampling points allocated at 250 m, 500 m and 1000 m from the platforms, totaling 18 samples for each platform. The determination of V, Mn, Co, Ni, Cu, Zn, Cd and Pb was performed by Inductively Coupled Plasma Mass Spectrometry (ICP-MS) after solid phase extraction using online and offline montages. Inductively Coupled Plasma Optical Emission Spectrometry (ICP-OES) was used to determine Ba, Fe, Zn, Cd, Ni, Cr, Cu, Ag and Al in produced water and Ba in seawater. Uranium and molybdenum determinations were performed by ICP-MS with an ultrasonic nebulizer. Using one liter of produced water and twenty liters of seawater, after radiochemical separation, ^{210}Pb was determined by beta counting of ^{210}Bi, and ^{226}Ra and ^{228}Ra were determined by gross alpha and beta counting rate using a proportional 10-channel low-level proportional counter. The element concentrations in the effluents were, in general, compatible with the literature values and below the maximum acceptable values established by the Brazilian legislation. The measured activity concentrations ranged from 0.8 to 6.0 Bq L^{-1} for ^{226}Ra, from 0.7 to 8.2 Bq L^{-1}

for ^{228}Ra and from 0.04 to 1.9 Bq L^{-1} for ^{210}Pb, similar to values reported for other produced water around the world. A strong correlation between Ba and radium isotopes in produced water was observed. Element concentrations in seawater were similar at the two study sites. The values obtained for ^{226}Ra, ^{228}Ra and ^{210}Pb in seawater, were below 5 mBq L^{-1}. The results showed a high degree of dilution regarding the produced water samples for barium, radium isotopes, copper, cadmium, nickel and ^{210}Pb. In spite of the elevated element and natural radionuclide concentrations in the produced water, concentrations in the seawater from the vicinity of the effluent discharges were at the local background level.

Chapter 3 – The chapter deals with the development of radiochemoecological approach for evaluation of the maximum allowable input (MAI) of radioactive and toxic substances to marine environments, which combines assessment the capability of seawater for self-purification and toxicity of the pollutants. The net rate of self-purification is proposed to estimate using the time-series data on concentration in seawater of the fallout radionuclides, both well-soluble (^{137}Cs, ^{90}Sr) and particle-reactive ($^{239+240}$Pu). This provides opportunity for direct evaluation of the levels of maximum allowable discharge into marine environments of nuclear and non-nuclear pollutants; for differentiation between their effective and ecological half-lives; for assessment of capacity of aquatic ecosystems in respect of accumulation of contaminants. Quantification of toxicity may be expressed, in general, as the classic values of the maximum allowable concentrations (MAC), but there is an uncertainty regarding the applicability of many of national and international levels of MAC for environmental regulation of anthropogenic impact in marine ecosystems. This is particularly important for comparative assessment of the influence of nuclear and non-nuclear contaminants having different nature of biological effects and different dimensions. An approach for such the comparison is the evaluation of dosimetric equivalents (e.g., Grey-equivalent), which are used to compare the effects of ionizing radiation and other stressors in aquatic ecosystems. But this would require every time to carry out the parallel ecotoxicological and radioecological dose-effect experiments that is not always accessible. In order to overcome this difficulty, a dimensionless value expressed as a fraction of maximum ability of marine organisms for resistance to radiation or toxic exposures is discussed in this chapter based upon the early reported experiments on the effects of mercury and incorporated radionuclides on the marine unicellular algae. The developed MAI criterion may be useful tool for scientific justification of environmental control of pollution of seawater with the nuclear and non-nuclear substances at a common methodological basis. The chapter highlights also the questions related to environmental assessment of the state of marine ecosystems, which have been subjected to the structural changes due to anthropogenic influence, and their response through biogeochemical processes of compensatory homeostasis.

Chapter 4 – When the Saint Lawrence Seaway macro-project was completed in 1959, it opened a "Fourth Coast" for the USA. (Its East, Gulf and West Coasts were provided by Nature.) Excavation of a Canal connecting the Gulf of California, through Mexico to near the city of Mexicali and beyond to the USA would permit seawater flooding of the below-sea level Salton Sea, a briny, contaminated and smelly endorheic lake that is fast disappearing after its unnatural creation by man-made flash flooding in 1905-07. This chapter will fathom some mysteries of the Salton Sea, including some speculative informed comments that might seem like science-fiction or science faux. However, the author takes the discussion tack that a practical rendition of the possible future geographical facts for this inhabited and cultivated desert region of southern California and northern Mexico, namely "Alternative Zero", is more real-world than Peter D. Ward's *The Flooded Earth: Our Future in a World Without Ice Caps* (2010). Seawater can be a beneficent liquid, sometimes tranquil, soft and tractable. And, sometimes, it causes peril for humans, as well as economic losses. Not insignificantly, the San Andreas fault—the most well-known plate tectonic boundary in the world—parallels the easternmost edge of the existing Salton Sea; the bi-national seawater Canal and Megaport constructors, influenced by Walt Disney's "Imagineering" urban planning concepts, must make certain macro-project design allowances recognizing that geophysical reality. For Mexico, international commercial trade and tourism are the only industries that allow a net flow of wealth from richer to poorer nations. "Alternative Zero" has the potential to alter the climate of the Southwest as well as northern Mexico for the better.

Chapter 5 – An evaluation of the SO_2 absorption in Caribbean seawater is measured by the addition of sulphuric acid at different conditions. The Caribbean is characterized by two main seasons: dry and hurricane seasons such that the variations nearshore currents are present. The measurements were carried out in closed systems, using hypothetical effluents where the nearshore currents were simulated. The results indicated that the addition of sulphuric acid at high speed current reduces the buffer capacity quickly and its acidity can increase until 6.6.

In: Seawater
Editor: Mathew R. White
ISBN: 978-1-62808-606-5

Chapter 1

IMPACT OF INTERNAL WAVE BREAKING ON THE OCEAN UPPER LAYER FORMATION

Alexander Yu. Benilov*
Applied Marine Physics Laboratory, Acute Solution,
Highlands, NJ, US

ABSTRACT

The physical models of air-sea interaction mainly concentrated on the processes at the air-sea interface, surface waves, mixing within the ocean upper layer and the formation its hydro-physical structure including its lower boundary. The last one is relatively narrow seawater layer with the sharp seawater density increase with depth growth. This layer is often called the pycnocline. The dynamical and thermo-dynamical contribution of the pycnocline into the upper layer is routinely ignored because the very strong density stratification makes the lower boundary impermeable for dynamical and thermo-dynamical interactions with the main ocean body below the pycnocline. The exception is only internal waves propagating at the pycnocline. However the internal waves, like the surface waves, break at randomly distributed locations at the pycnocline generating turbulent patches which violate the pycnocline integrity and so its impermeability. The paper describes the effect of internal wave breaking turbulence on the structure of the hydro-physical and biogenic fields near the ocean surface. The exchange of turbulent

* Corresponding author: Tel.: 1 (732) 291 0253; e-mail: <abenilov@rocketmail.com>.

energy, heat and biogenic salts between the ocean upper layer and the main body of the ocean below is caused through the random patches of active turbulence within zones of internal wave breaking. The variability of the upper layer temperature, concentration of biogenic salts and the turbulent energy is caused by the turbulence of the breaking zones, its variability and the breaking zones statistics. The turbulent horizontal diffusion of the ocean upper layer spread the non-uniformities of these fields into larger horizontal scales than the breaking zones scales. Estimates of the manifestation of the discussed mechanism are shown in the mean values and variances of the surface non-uniformities of the upper layer fields as well in their spectrums. The simple model examples of the field of internal wave breaking turbulence allow quantitative evaluation.

1. Introduction: The Problem Formulation

Remote methods of measuring hydrophysical fields on the ocean surface have enjoyed extensive development during last three decades [Stewart, 1985; Martin, 2011]. This technique is in remote sensing applications since 1970's to record surface elevations [Zagorodnikov, 1978; Glazman, 1991], and surface temperature [Galaziy, Shifrin and Sherstyankin, 1979; Minnett, 1990; Schluessel and et al., 1990], wind above the ocean surface [Kelly and Caruso, 1990; Lalbeharry, Khandekar, Peteherych,1990; Willard and Pierson, 1990] and biological surface fields [Dickey, 1991; Hood and et al., 1990]. The remote methods are of fundamental interest to investigation of the ocean using artificial earth satellites [Fedorov and Sklyarov, 1980; Stewart, 1985; Dickey, 1991; Wahl and Simpson, 1990; Martin, 2011].

Among the diverse processes responsible for forming surface non-uniformities in hydrophysical fields, internal waves, which are encountered everywhere in the ocean, play an important role. They influence surface temperature, and they change the nature of surface disturbance, forming slicks--areas of smoothing of small-scale wave components [Fairbridge, 1966; Steele, Thorpe and Turekian, 2009].

The action of internal waves upon surface characteristics of hydrophysical fields may manifest itself in two ways. First, the internal wave field induces a field of surface movements which can lead to formation of surface non-uniformities in hydrophysical fields of the scale of internal waves [Monin and Piterbarg, 1977; Fedorov, 1981]. Second, internal waves affect the turbulence of the ocean upper quasi-uniform layer. The turbulence is responsible for vertical mixing in the upper layer, and in turn it has a strong influence upon

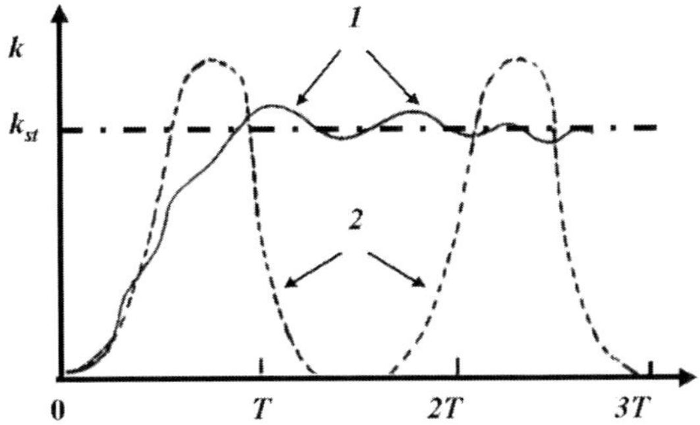

Figure 1. An example of the internal wave effect on the turbulent kinetic energy k (TKE) in a current of stably stratified fluid [Ivanov et al., 1982]; 1 - TKE modulated by the internal wave with the period T near a steady state k_{st} induced by the mean shear, 2 - local generation of turbulence by the internal wave shear.

surface and internal waves [Hasselmann, 1966; Fedorov and Sklyarov, 1980; Ivanov et al., 1982; Benilov, 2013]. The effect of internal waves on upper layer turbulence manifests itself in the form of modulation of the turbulence statistical characteristics by internal waves [Ostrovskiy and Soustova, 1977; Ivanov et al., 1982; Ostrovskiy and Piterbarg, 1997], and as an effect of turbulence, which is produced by the internal waves breaking, upon the upper turbulent layer [Barenblatt and Benilov, 1982; Benilov 1984, 1986; Benilov and Gavrilin, 1988; Bock et al., 1990; Moum et al., 1990; Wijesekera and Willon, 1990]. Figure 1 shows an example calculated by Ivanov et al.[1982] that demonstrates the internal waves effect on turbulence. In the case of weak interaction (line 1) the TKE (turbulent kinetic energy) is modulated by the shear of internal wave with the period T near a steady state k_{st} induced by the shear of mean current and the oscillations of TKE have the period of the internal wave. The events of local generation of turbulence by the shear of internal wave are illustrated by line 2.

The effect of turbulence on surface waves differs fundamentally for waves of scale ℓ greater than 10 cm (gravity waves), and for waves of scale ℓ less than or equal to 10 cm (ripples). In the first case ($\ell > 10\ cm$) the propagation of surface gravity waves obeys well to the potential theory. The upper layer turbulence occurs due to the energy flux $\boldsymbol{j}$ from the potential wave motion into the vortex turbulent motion. This energy flux transfers the kinetic energy E_φ of the wave field into the turbulent kinetic energy [Benilov, 1973; Benilov and Ly, 2002] in the direction opposite to the gradient of E_φ that is:

$$\boldsymbol{j} = -\nu_T \boldsymbol{grad} E_{\boldsymbol{\varphi}} \tag{1}$$

where ν_T is the coefficient of turbulent exchange of wave energy E_φ. This coefficient may be assumed to be proportional to the coefficients of exchange of momentum and turbulent energy. The wave energy spreads throughout the entire upper layer by the energy flux $\boldsymbol{j}$. This is what leads to smoothing the surface waves with scale $\ell > 10$ cm. Also, the contribution of the internal waves should be taken for consideration. At the bottom of the upper mixed layer there are internal waves, which propagate within the density discontinuity layer and some time break, and then localized turbulent zones occur in the vicinity of the lower boundary of the upper layer. These turbulent zones may widen, raising the intensity of turbulence in the upper layer. It means that the upper layer may have an additional influx of turbulent energy from the area of the internal wave breaking. As a result of this, the coefficient of turbulent diffusivity ν_T increases, and because ***grad*** E_φ is not small near the free surface, the turbulent flux of wave energy increases. Hence, it follows that areas of smoothing on the free surface will appear over areas of breaking internal waves. In the case of low turbulence intensities within breaking zones the turbulent energy flux will have the reverse direction, i.e. the upper layer turbulence will feed the turbulence of the breaking zone. As a consequence the smoothing of the ocean surface is expected in the areas unperturbed by the internal wave breaking, since the level of unperturbed turbulence would turn out to be higher than the level of perturbed turbulence.

Fluctuations of the ocean surface having a scale range of $\ell \leq 10$ cm (ripples) are different in nature. A detailed analysis of experimental data on wind waves carried out by Zaslavskiy [1981] and Zaslavskiy and Leykin [1981] shows significant deviations of the measurements from the conclusion of linear potential theory concerned with this range of scales. Yefimov and Khristoforov [1971] noted that these differences indicate that the velocity field is determined to a significant extent by turbulent motion in this range of scales. The study [Zaslavskiy, 1981; Zaslavskiy and Leykin, 1981] of the properties of the empirical dispersion relation of small-scale waves obtained by processing the observation data confirms the conclusions [Yefimov and Khristoforov, 1971]. It follows from the empirical fact that the scattering magnitude $\Delta\omega$ of the empirical dispersion relation is about the same order of magnitude as the frequency ω itself. As one can see, the single-valued connection between both the frequency ω and the wave vector **k** does not take place here, and it is one of the principal indicators of the turbulent nature of fluid motions. The random vortex nature of surface turbulence may be a product of irregularities of tangent stresses generated by the atmospheric turbulent boundary layer on the ocean surface and, also, the breaking of surface waves. According to the data [Kuznetsov, 1978], the variations of the friction velocity v_* of the atmospheric boundary layer correlate well with the r.m.s. of the sea surface elevations measured in the frequency range 1-10 Hz.

Relying on the experimental facts presented above, we can identify small-scale fluctuations in the ocean surface with ordinary turbulence, which may be described, for instance, within the framework of the modern semi-empirical turbulence theory. What are the consequences of this assumption for the ocean surface? Modulation of the turbulence of the upper layer by internal waves leads to the appearance of alternating spots or slicks. The scale of this phenomenon will be equivalent to that of internal waves. Non-uniformities caused by the breaking of internal waves are superimposed over this pattern. For the range of scales of surface fluctuations under examination here ($\ell \leq 10$ cm), the effects will be directly opposite to what was noted in the case of fluctuations with scale $\ell \geq 10$ cm. When the intensity of upper layer turbulence is increased by an influx of turbulent energy from zones of the internal wave breaking, the surface would be less smooth than in areas which are not perturbed by breaking waves, while on the other hand the surface would be smoother wherever upper layer turbulence feeds the turbulence of the breaking zone and wherever a drop in the level of turbulence within this layer exists. In the case where the turbulence of the center of the breaking zone significantly exceeds the turbulence in the upper layer, one breaking zone projects itself to the surface in the form of two smoothing spots, which merge into one as the turbulence of the breaking zone drops with time. This double-relationship pattern stems from nonuniformities in the distribution of turbulence within the breaking zone itself - the turbulent intensity is high in its center, and low on the periphery. This paper is devoted to the discussion of the latter mechanism.

To reach the posed objective we need to build a model satisfactorily describing formation of the upper mixed layer, the equilibrium characteristics of which are determined by atmospheric influences upon the surface and by background stratification outside the layer. The model must describe interaction of upper layer turbulence with abyssal turbulence generated by the breaking of internal waves and by other mechanisms. It seems as if the theory of this phenomenon could be derived using the traditional methods of the semi- empirical theory of turbulence. Below we shall use the approaches developed in [Barenblatt and Benilov, 1982; Benilov 1984, 1986] to fulfill the listed requirements.

2. The Upper Turbulent Layer

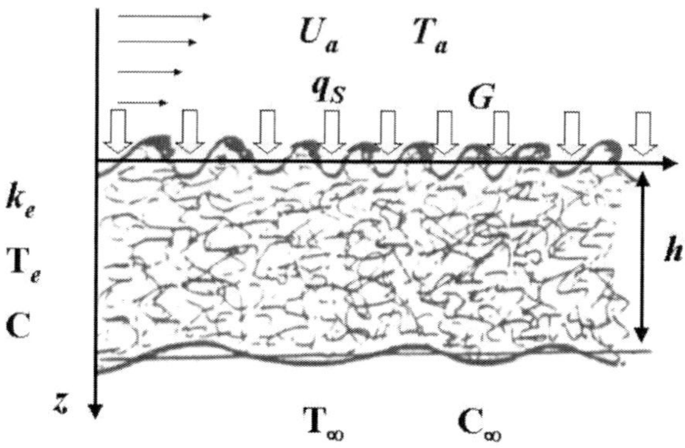

Figure 2. The oceanic turbulent upper layer in thermo – dynamic equilibrium with the atmospheric boundary layer without the internal wave breaking turbulence; U_a is the wind velocity; T_a is the air temperature; q_s is the energy flux; G is the heat flux; h is the upper layer thickness; k_e, T_e, C are the TKE , the temperature and the biogenic salts (mineral nutrition) concentration of the upper layer; T_∞, C_∞ are the temperature and the biogenic salts (mineral nutrition) concentration of the main water body of the ocean.

We will examine, as a model of the upper turbulent layer of the ocean, a liquid occupying a region of space $z \geq 0$ (Figure 2, axis z is directed vertically downward). Density stratification is determined by vertical change in temperature, T. The analysis is conducted on the basis of an approximation of the upper quasi-uniform layer continually filled with turbulence and separated from the non turbulent region by a sharp boundary [Barenblatt, 1977] (mixing ceases where turbulent kinetic energy, k, reduces to zero). Thus it is assumed that the upper mixed layer occupies the layer of the thickness h ($h \geq z \geq 0$), the temperature in the upper layer does not depend on the vertical coordinate z, but that it does depend on time t: $T = T_s(t)$ (the traditional approximation in models of the upper quasi-uniform layer [Kraus and Businger, 1995]), and that beyond its lower boundary the temperature is constant and equal to $T = T_\infty = const.$ The state of the atmosphere over the ocean is characterized by wind velocity U_a and temperature T_a at a standard measuring height $z = z_a < 0$. The wind above the free surface of the ocean generates the wind surface waves. The surface wave breaking produces fluxes of turbulent energy q_s and as well as momentum [Benilov, 2013] that maintains the turbulence of the upper layer (Figure 2.). According to the estimates carried out in [Longuet-Higgens, 1969; Benilov, 1973; Benilov and Ly, 2002] these fluxes are associated with the phase velocity C_0 of the spectral maximum of wind waves. The energy flux is given by the expression $q_s \approx 5\times10^{-8}C_0^3$. In the case of developed waves the phase velocity C_0 of the spectral maximum satisfies the approximate equality $C_0 \approx U_a$ [Kitaygorodskiy, 1970]. Then, in the case of developed waves, the energy flux q_s becomes

$$q_s \approx 5\times10^{-8}U_a^3, \tag{2}$$

that corresponds to the steady situation depicted in the diagram in Figure 2.

Heat flux G from the atmosphere to the upper layer may be represented in the form

$$G = C_T\rho_a c_{pa}U_a(T_a - T_s), \tag{3}$$

where $C_T \approx 10^{-3}$ is the heat transfer coefficient [Kitaygorodskiy, 1970; Kanta and Clayson, 2000], ρ_a is the air density, and c_{pa} is the specific heat of air. In the absence of breaking of internal waves, which in this approximation are excited by turbulence of the upper layer at its lower boundary $z = h$, the atmosphere and the upper layer are in thermal and dynamic equilibrium: the energy flux from the surface waves is completely expended by dissipation and work against the buoyancy forces.

When the atmosphere and upper layer are in thermal equilibrium, $G = 0$ and $T_a = T_s$. Thus the equilibrium of the upper layer in steady state is determined completely by three constant dimensional parameters q_s, $\delta T = T_a - T_\infty$, and the buoyancy parameter, $\beta = g\alpha$, where α is the coefficient of thermal expansion of liquid and g is the acceleration due to gravity. Therefore:

$$[q_s] = L^3\tau^{-3};\ [\delta T] = \theta;\ [\beta] = L\tau^{-2}\theta^{-1} \tag{4}$$

where L is the length dimension, τ is the time dimension, θ is the temperature dimension.

The effect of the mean velocity shear is not accounted in this approach because, by the initial assumption, the energy flux from the surface waves is the dominant energy source. It should be noted that a shear free approximation provides a realistic description of upper layer turbulence (see for example [Kitaygorodskiy, 1970; Benilov, 1973; Benilov and Ly, 2002]), and for the purposes of the problem under examination it appears sufficient.

Using the characteristic parameters (4), we can write the integral energy balance of the upper layer assuming that the integral dissipation of turbulent energy is proportional to its production, as governed by the energy flux q_s. This estimate of dissipation is broadly employed in simulation of the upper quasi-uniform layer in the approximation under examination here [Kraus and Businger, 1995]. The energy balance equation has the form

$$k_e = \kappa_1 q_s^{2/3}, \; q_s = \kappa_2 g\alpha(T_e - T_\infty)h\sqrt{k_e}, \; T_e = T_a, \tag{5}$$

where k_e is the integral TKE of the upper layer in the steady thermo - dynamic equilibrium state; κ_1, κ_2 are dimensionless empirical constants. Note that the definition of k_e in Equation (5) is obtained uniquely by the analysis of the dimensions of the characteristic parameters (4). Equation (5) makes it possible to determine the dependence of the thickness of the upper layer h on external parameters:

$$h = \kappa_h \frac{q_s^{2/3}}{\beta(T_a - T_\infty)}, \; \kappa_h = \frac{1}{\kappa_2\sqrt{\kappa_1}} \tag{6}$$

Equation (6) provides a correct qualitative dependence of the thickness of the upper quasi-uniform layer on the intensity of wind action upon the ocean surface and on the bulk temperature difference $\delta T = T_a - T_\infty$. This equation determines the thickness of the upper layer on the condition that the thickness of the Ekman layer $L_e >> h$. Kitaygorodskiy [1970] gives the equation which is similar to (6) and where h is the thickness of the layer caused by wind mixing. Equations (2) and (6) allows determination of the thickness of the upper layer h using atmospheric parameters U_a and T_a and the temperature of ocean water T_∞ beyond the upper layer, if the numerical value of the empirical parameter κ_h is known. Dimensionless coefficients κ_h, κ_1, κ_2 may be determined by assuming typical values for U_a, $\delta T = T_a - T_\infty$, k_e, h and α [Kitaygorodskiy, 1970; Kraus and Businger, 1995; Monin and Ozmidov, 1985]: 10_3 cm/sec; 10°C; 1 cm^2/sec^2; 3×10^3 cm; $10^{-4}(°C)^{-1}$, respectively. Then we obtain the following estimates:

$$\kappa_h \approx 2\times10^2,\ \kappa_1 \approx 7\times10^{-2},\ \kappa_2 \approx 2\times10^{-2}. \tag{7}$$

3. Appearance and Development of Turbulence in the Presence of a Breaking Internal Wave

Instability of internal waves and their breaking represent the principal mechanism of generation of small-scale turbulence within the pycnocline, where the greatest density of internal wave energy is concentrated [Monin and Piterbarg, 1977]. The largest proportion of this energy is usually related to the lowest mode of internal waves, such that the adopted simplified assumption of a two-layer liquid is fully justified. The frequency of the internal wave, ω, is associated in this case with the wave number k by the relationship:

$$\omega^2 = \frac{g\alpha(T_a - T_\infty)k}{1+\coth(kh)}, \tag{8}$$

which in real conditions would be valid for the lowest mode of internal waves. In the case of short waves ($kh >> 1$), Equation (8) gives us the asymptotic relationship $\omega^2 = g\alpha(T_a - T_\infty)k/2$, which is similar to the dispersion relation for surface waves over deep water. In the case of long waves, we obtain the following relationship from Equations (8) and (6):

$$\omega = \sqrt{\kappa_h q_s^{1/3}}\,k. \tag{9}$$

According to this expression the phase velocity of the long waves C_p is completely determined by the energy flux from surface waves q_s, and is equal to $C_p = \sqrt{\kappa_h \sqrt[3]{q_s}} \approx 10\sqrt{2\sqrt[3]{q_s}}$ for the typical conditions assumed earlier. Finally we obtain $C_p \approx 50 cm/s$ that gives the period $\tau = 2\pi/\omega \approx 6\times10^2 s$ for a wave of the wave length $\lambda = 2\pi/k \approx 3\times10^4 cm$; waves with such spatial-temporal scales are typical in the ocean.

Let us assume furthermore that internal waves excited by upper layer turbulence at the lower boundary of the upper layer $z = h$, break, transforming their energy into the energy of a turbulent spot (Figure 3). In the case of plane internal wave the turbulent breaking zone has the shape of a cylinder (Figure 4, (2)) with the generatrix which is perpendicular to the direction of wave propagation. This type of breaking zone was observed in laboratory experiments in which internal waves were generated at a density discontinuity by a velocity shear [Browand and Winant, 1973]. According to the theory on dynamics of such isolated turbulent formations [Benilov and Gavrilin, 1988] and laboratory experiments [Schooley and Stewart, 1963; Browand and Winant, 1973] the evolution of such turbulent zones in time proceeds during the initial stages in the same way as in a liquid of uniform density. Thus, in the initial stage of evolution of the breaking zone, the approximation of a non-stratified liquid is fully permissible. In this stage the TKE of the internal wave breaking zone k_w is assumed to be significantly greater than the TKE in the upper layer k_S. Then the distribution of k_w within the breaking zone would be approximately axisymmetrical (the axis of symmetry in the plane (z, y) has the coordinates $(z = h,\ y = 0)$, where y is the transverse coordinate parallel to the direction of wave propagation), and the influence of upper layer turbulence upon the turbulence of the breaking zone may be ignored. In later stages, when k_w and k_s are commensurate and $k_s >> k_w$, the influence of the upper layer on k_w would now be significant. However, stratification begins to play an important role in this case, and the energy flux from the upper layer into the breaking zone would be expended on work against the buoyancy force. Keeping in mind this circumstance, we may assume, approximately, that within the breaking zone, the influx of turbulent energy from the upper layer is completely expended on work against Archimedean forces. This assumption allows us to use the approximation of an non-stratified liquid to analyze the turbulence in the breaking zone, considering that the energy fluxes across the boundary of this region are zero. The diagram in Figure 5 shows how the internal wave breaking turbulence disturbs the thermo – dynamic equilibrium and opens mass exchange with the main ocean body through the breaking zones. It should be noted that when we evaluate changes in upper layer turbulence elicited by interaction with the turbulence of breaking waves, the exchange of energy between them must be accounted for in all stages of evolution of the breaking zone: this exchange of energy is precisely what leads to change in magnitude of k_s near the breaking zone. It is clear from this line of reasoning that development of a turbulent region proceeds differently within the upper layer ($z < h$) and outside it ($z > h$): When $z < h$, $k_w \neq 0$ at the boundary of contact between the upper layer and the breaking zone, while when $z > h$, the liquid is unperturbed outside the breaking zone, turbulence is absent, and k_w is equal to zero at the boundary of the turbulent zone. A symmetrical description of the turbulence of the breaking zone in the case

$z < h$, such that $k_w(z,y,t) = k_w(z,-y,t)$, is the simplest approximation simulating such asymmetry of the breaking zone and satisfying the conditions enumerated above.

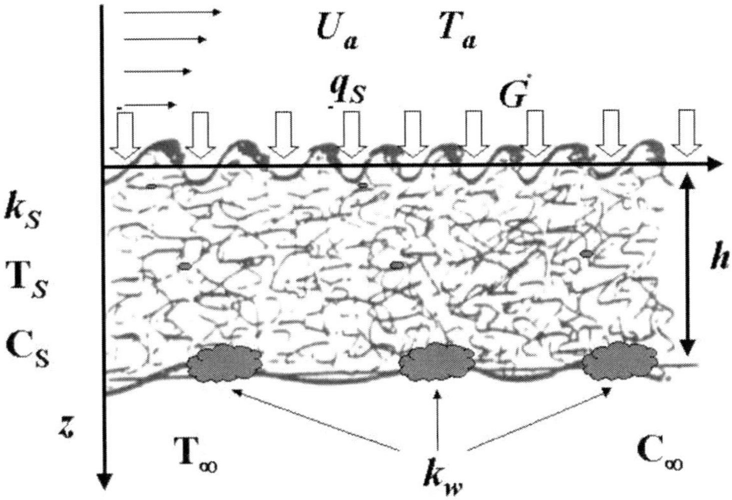

Figure 3. The oceanic turbulent upper layer in thermo – dynamic equilibrium with the atmospheric boundary layer in the presence of the internal wave breaking turbulence. There is no interaction between the upper layer turbulence and the internal wave braking turbulence.

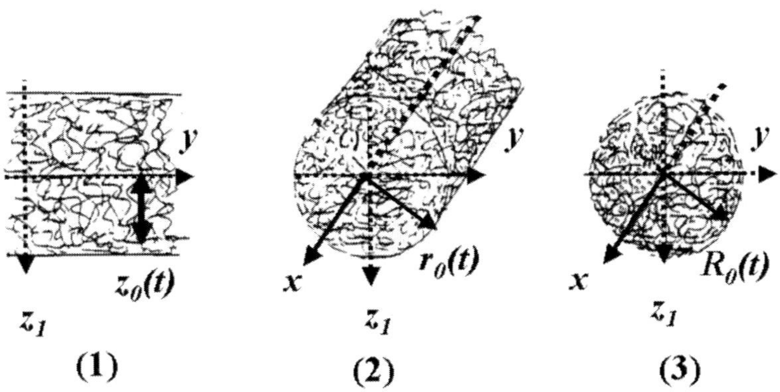

Figure 4. The different approximations of the types of turbulent regions caused by the internal wave breaking; 1-the flat turbulent layer, 2 – the cylindrical turbulent region, 3 – the spherical turbulent region.

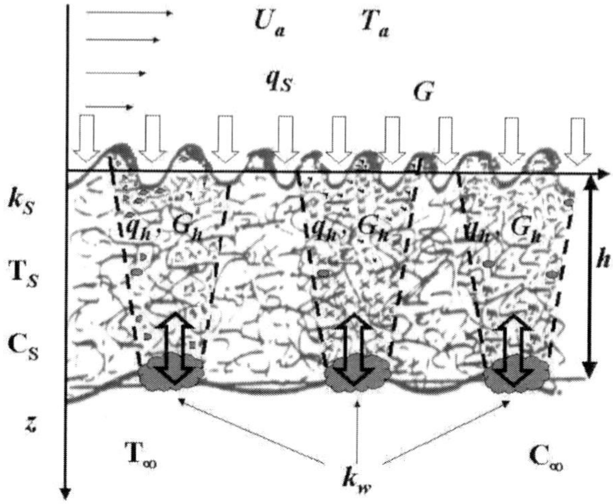

Figure 5. The diagram of interaction of the turbulent upper layer with the internal wave breaking turbulence.

The boundary of contact with the upper layer is situated at depth $z = h$ at the boundary with the non-turbulent region of the seawater where $k_w = 0$; however, when $z = h$ the vertical energy flux must be equal to zero and $k_w \neq 0$. Thus, this approximate description of the evolution of a turbulent region made by internal wave breaking and its interaction with the upper layer may be reduced, in the extremely rough approximation under examination here, to studying development of a turbulent region within a liquid of uniform density in time. The region itself is delimited from the upper layer by the plane $z = h$ and from the non-turbulent liquid by a certain surface which, ignoring collapse of the turbulent region in response to stratification and subsequent viscous spreading [Barenblatt,1978; Benilov 1989], assumes one simple form or another soon after the region forms. The energy fluxes at the boundary of the region are always equal to zero while the turbulent kinetic energy is equal to zero only at the boundary with the non-turbulent region. These requirements are fully satisfied by a model of evolution of turbulent regions in a liquid of uniform density that is initially unperturbed, since the fact that the energy flux at boundary $z = h$ is equal to zero permits us to supplement the definition of the breaking zone as being symmetrical within the range of z values less than h, so that the turbulent region would in fact satisfy the requirements enumerated above for the case $z > h$.

3.1. Flat Turbulent Layer

To simulate local interaction of the upper layer with turbulence created by a breaking internal wave, we will examine the following simplest problem. The breaking of an internal wave generates an initial turbulent layer of thickness $2a$ (Figure 4, (1)) and having an initial charge of turbulent energy I_1 (per unit area). The thickness of the turbulent layer $2a$ is significantly less than its horizontal dimensions. We will not examine below the influence of shear explicitly. However, we should consider that the work of Reynolds stresses on the mean velocity shear makes a contribution into the magnitude of I_1.

Following to [Barenblatt, 1982], we will examine the evolution of a flat turbulent region that has arisen from an instantaneous source (a breaking internal wave) which initially occupies a horizontal flat layer of small thickness at depth $z = h$. The evolution of the initial turbulent region in terms of changes of the turbulent kinetic energy k_w in time and in space may be described within the framework of the differential semi-empirical theory of turbulence [Monin and Yaglom, 1987] in a shear-free approximation, where self- similarity and kolmogorov's hypotheses are used for the purposes of closure. In the case of a flat turbulent layer (Figure 2c. 1), k_w depends only on z and t, and the turbulent kinetic energy budget equation, written with variables $z_1 = z - h$ and t, takes the form

$$\partial_t k_w = \gamma_1 \partial_{z_1} \ell \sqrt{k_w} \partial_{z_1} k_w - \gamma_2 k_w^{3/2} / \ell, \tag{10}$$

where γ_1, γ_2 are empirical constants, the values of which were determined by comparing calculations with experimental data [Frost and Moulden, 1977; Hoffman, 1989], and ℓ is the turbulence mixing scale, which will be defined below. Equation (10) is close in type to the equations of nonlinear heat conductivity and to a similar equation describing the spread of a turbulent region from a constantly acting source on the surface [Barenblatt, 1977; Barenblatt, 1978; Benilov, 1984]. These equations are unique in that the rate of advance of the boundary of the turbulent region within the non-turbulent liquid is finite. Therefore the boundary conditions requiring reduction of turbulent energy and its flux to zero may be applied to this moving boundary, subject to definition. In the case under examination here, this boundary is a plane situated a distance z_0 (t) from the lower boundary of the upper layer. Final closure of Equation (10) requires assigning a value to the turbulence scale ℓ . The simplest closure hypothesis of [Barenblatt, 1977; Barenblatt, 1978] states that the scale ℓ is some fixed proportion of the dimension of the turbulent region and then that is, $\ell = \gamma_\ell z_0(t)$, where γ_ℓ is a proportionality constant.

Equation (10) is supplemented by an initial condition. Having in mind symmetrical continuation into the interior of the upper layer, it is assumed that the function determining the initial distribution of turbulent energy is symmetrical relative to the middle plane of the breaking zone ($-a < z_1 < a$), and that it is equal to zero outside the breaking zone, where $2a$ is the initial thickness of the turbulent breaking zone. The initial condition can be conveniently written in the form

$$k_w(z_1,0) = \frac{I_1}{a} u_1(\tau_1),$$

$$\tau_1 = \frac{z_1}{a},\ I_1 = \int_{-a}^{a} k_w(z_1,0)dz_1,\ \int_{-1}^{1} u_1(\tau_1)d\tau_1 = 1, \tag{11}$$

where $u_1(\tau_1)$ is an even dimensionless function of its one dimensionless argument, equal to zero when $\tau_1 \geq 1$ and I_1 is the total charge of turbulent energy generated by breaking of the internal wave. Thus the problem under examination here has been reduced to solving a Cauchy problem for Equation (10) within the domain $-z_0(t) \leq z_1 \leq z_0(t)$, given initial condition (11). The function $z_0(t)$ is the thickness of the breaking zone at time t, that is previously unknown and must be determined in the course of solving the problem as well.

The solution to such a problem is not a self - similar one. It depends on parameters I_1, t, z_1, γ_1, γ_2, and a, of which z_1 and a have identical dimensionality. But what is of fundamental interest here is the asymptotic behavior of the solution for the case of a layer having a thickness much greater than the initial thickness ($z_0(t) \gg a$), since the influence of the details of the initial conditions should disappear in this case. We will therefore assume that incomplete self-modeling exists in relation to the problem under examination here, and consequently the following form of the solution is valid:

$$k_w(z_1,t) = \frac{4A_1^2 B_1(\xi_1)}{[(3+2\nu_1)\gamma_1\gamma_\ell]^2 t^{2n_1}},\ \xi_1 = z_1/z_0(t),\ \ z_0(t) = A_1 t^{1-n_1}, \tag{12}$$

$$A_1 = \zeta_0 I_1^{(1-n_1)/2} a^{(3n_1-1)/2},\ n_1 = \frac{1+2\nu_1}{1+3\nu_1},$$

where A_1 and ζ_0 are constants. Parameter ν_1 characterizes the difference of the exponential law of the layer's expansion in the case of incomplete self-similarity from the case of complete self-similarity, which exists when γ_2

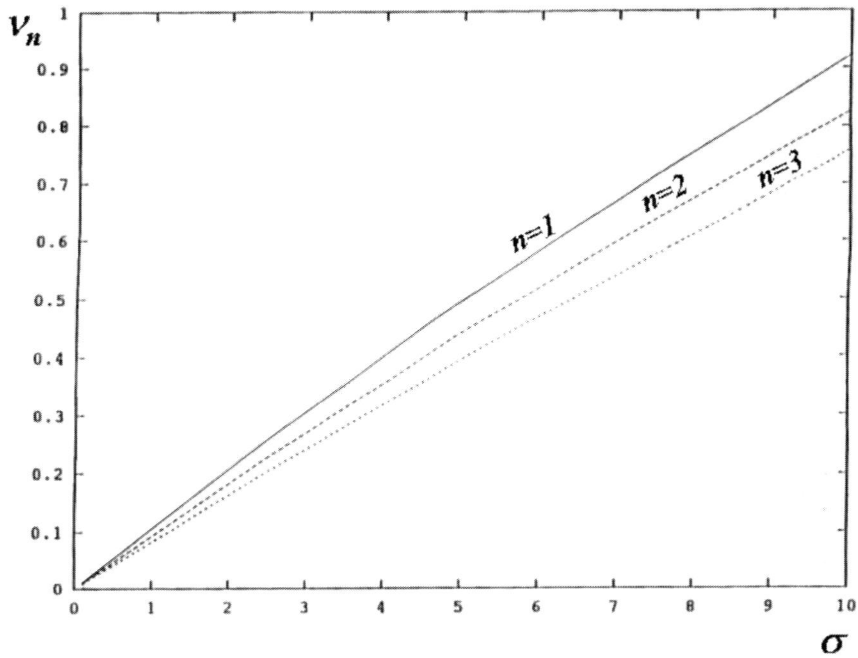

Figure 6. The eigenvalue ν_n (n = 1, 2, 3) as a function of the parameter σ; 1- flat turbulent layer, 2 – cylindrical turbulent region, 3 – spherical turbulent region.

= 0, that is, when dissipation equals zero and the initial energy of the charge of turbulence created by the breaking wave I_1 persists at all times t; in this case ν_1 = 0. Substituting (12) in Equation (10), for function B_1 (ξ_1) we get an ordinary differential equation ($\xi_1 = z_1 / z_0\,(t)$):

$$d_{\xi_1}\sqrt{B_1}d_{\xi_1}B_1 + \xi_1 d_{\xi_1}B_1 + (1+2\nu_1)B_1 - \sigma B_1^{3/2} = 0\,,\ \sigma = \frac{\gamma_2}{\gamma_1\gamma_\ell^2}. \qquad (13)$$

In view of the symmetry of the problem, B_1 (ξ_1) = B_1 ($-\xi_1$), we can limit ourselves to deriving a solution only for the domain $0 \le \xi_1 \le 1$. Thus we seek a nontrivial solution to Equation (13) in the presence of the conditions:

$$d_{\xi_1}B_1(0) = d_{\xi_1}B_1^{\frac{3}{2}}(1) = B_1(1) = 0. \tag{14}$$

This brings us to a classical situation of a nonlinear problem involving eigenvalues. We must arrive at a nontrivial solution to Equation (13) in the presence of conditions (14) and determine the value of parameter $\nu_1(\sigma)$ at which this solution exists. Figure 6 shows the eigenvalue ν_1 given as a function of σ, ν_1 increases from 0 to about 0.9 while σ increases from 0 to 10. Figure 7(a-d, curves $n = 1$) shows the solution B_1 (ξ_1) for the cases $\sigma = 1.25, 2.5, 5.0, 10.0$. Integration of Equation (13) with respect to the entire thickness of the turbulent breaking zone shows that when parameter σ assumes arbitrary positive values, ν_1 is positive ($\nu_1 = 0$ when $\sigma = 0$ and increases with increasing σ, such that n_1 in (12) is finite with respect to all σ and lies within the limits $1/3 < n_1 < 1$). Hence it follows that if the initial thickness, $2a$, of the turbulent layer made by a breaking wave approaches zero, and we keep requirement that the asymptotic of the solution is invariable in the presence of long time intervals, but we could not keep the intensity of the turbulence charge I_1 constant because the presence of dissipation. It would have to grow, and it would have to do so in such a way that the moment

$$M_1 = I_1 a^{2\nu_1} \tag{15}$$

would remain constant. It is easy to see that the value of this moment remains constant for this solution for all values of time, and consequently, when dissipation is present (γ_2, $\sigma > 0$), the moment M_1 may serve as a measure of the intensity with which an internal wave breaks [Benilov and Gavrilin, 1988]. The parameter ν_1, the eigenvalue of the nonlinear problem examined here, is a characteristic of the internal wave breaking intensity as it follows of the equation (15). The following expansion is valid in relation to the solution of $B_1(\xi_1)$ in the vicinity of the boundary with the non-turbulent region ($\xi_1 \to 1$):

$$B_1(\xi_1) = \left(\frac{1-\xi_1}{2}\right)^2 - 2\,\frac{3+2\nu_1}{3}\left(\frac{1-\xi_1}{2}\right)^3 + \cdots \tag{16}$$

As one can see of the obtained solution, the influence of dissipation in the presence of a fixed value of the parameter σ upon the local structure of turbulence near the boundary of the turbulent region is small. The dissipation (or σ) begins making a contribution into the solution only by the third-order term containing the variable $(1 - \xi_1)/2$, because the function $\nu_1 = \nu_1(\sigma)$, shown in Figure 6 (curve $\nu = 1$, $0 \leq \sigma \leq 10$), is a monotonous growing function of s that is $\nu_1 = \nu_1(0) = 0$ and well follows to the approximation $\nu_1(\sigma) \approx 0.1083\sigma - 0.001633\sigma^2$.

Within the framework of the approximation under examination here, according to which all processes occurring in the upper layer are fully determined by external parameters (4), using dimensional analysis we can come up with a formula associating the mean moment $\overline{M_1}$ with these parameters. As a result of such analysis we get

$$\overline{M_1} = \chi_1 \frac{q_s^{\frac{2}{3}(2+2\nu_1)}}{(\beta\delta T)^{1+2\nu_1}}, \tag{17}$$

where χ_1 is an empirical dimensionless constant. According to expression (17) the intensity of wave breaking increases with growth of the wind velocity (q_s increases as U_a^3) and drops with growth the stratification (the temperature difference $\delta T = T_a - T_\infty$ increases) which is qualitatively correct from the standpoint of the physical possibility of the mechanism under consideration. Constants χ_1 in (17) may be evaluated using the same values for the characteristic parameters χ_1 is an empirical dimensionless constant. Constants χ_1 in (17) may be evaluated using the same values for the characteristic parameters that were used before, adopting additionally as an estimate of k_w, its greatest observed value in the thermocline $k_w \approx k_{wmax} \approx 10^{-4}\ m^2/s^2$, and assuming that the zone of breaking of the internal wave has an initial dimension on the order of the wave amplitude of about 3 m. Then we can adopt the estimate $a \approx 3\ m$. Using these numerical values of the parameters in expressions (11), (15), (17) we get the estimate $\chi_1 \approx 7.4$ in the case $\nu_1 \approx 0.257$ ($\sigma = 2.5$).

3.2. Cylindrical Turbulent Region

In the case of a flat system of internal waves propagating along the lower boundary of the upper layer in the y-direction, the turbulence within the breaking zone would be horizontally non-uniform along the y-axis. Before they join with the neighboring breaking zones, each zone would have a finite dimension. Within such zone, the turbulent kinetic energy decreases from its largest values at the center to zero at the boundary with the non-turbulent seawater. Given the assumptions made above, such non-uniform breaking zone may be approximated by an axisymmetrical, unsteady model of development of a certain circular cylindrical turbulent region with the axis of symmetry located at the depth $z = h$ (Figure 4,(2)). In this case the breaking zone would be represented by points in a region for which $0 \leq r \leq r_0(t)$ and $z > h$, where $r = \sqrt{y^2 + z_1^2}$ and $r = r_0(t)$ is the boundary of the breaking zone with the non-turbulent seawater (Figure 4, (2)). The turbulent kinetic energy and the turbulent energy flux are equal to zero at the boundary with the non-turbulent region. In view of symmetry, the turbulent energy flux is equal to zero and $k_w \neq 0$ at the boundary of contact with the upper layer when $z = h$.

Considering axial symmetry and assuming that as before, the turbulence scale is a certain fixed proportion of the dimension of the turbulent region, $\ell = \gamma_\ell r_0(t)$, the turbulent energy balance equation for k_w written in variables r and t has the form

$$\partial_t k_w = \gamma_1 \gamma_\ell \frac{r_0(t)}{r} \partial_r r \sqrt{k_w} \partial_r k_w - \frac{\gamma_2}{\gamma_\ell} \frac{k_w^{\frac{3}{2}}}{r_0(t)}, \tag{18}$$

Equation (18) is supplemented by an initial condition. Assuming that the initial distribution of turbulent energy is axisymmetrical and equal to zero outside a certain circular cylindrical region $0 \leq r < a$, where a is the initial radius of the turbulent region, we can write it in the form

$$k_w(r,0) = \frac{I_2}{2\pi a^2} u_2(\tau_2), \qquad \tau_2 = \frac{r}{a},$$
$$I_2 = 2\pi \int_0^a k_w(r,0) r dr, \ \int_0^1 u_2(\tau_2) \tau_2 d\tau_2 = 1, \tag{19}$$

where $u_2(\tau_2)$ is a dimensionless function of its one dimensionless argument, equal to zero when $\tau_2 \geq 1$; and I_2 is the total charge of turbulent energy per unit length of a cylindrical region with radius a, generated by a breaking internal wave. Thus the problem under investigation here reduces to solving a problem completely similar to the case examined in the previous section. Making the assumption of incomplete self-similarity as before, we get an asymptotic solution of the following form:

$$k_w(r,t) = \frac{A_2^2 B_2(\xi_2)}{[(2+\nu_2)\gamma_1\gamma_\ell]^2 t^{2n_2}},$$
$$\xi_2 = r/r_0(t), \ \ r_0(t) = A_2 t^{1-n_2}, \ n_2 = \frac{1+\nu_2}{2+\nu_2}, \tag{20}$$

where A_2 is a constant and the parameter $\nu_2(\sigma)$ is the eigenvalue of the new nonlinear problem. Substituting expressions (19) into Equation (18), we obtain an ordinary differential equation for function $B_2(\xi_2)$:

$$\xi_2^{-1} d_{\xi_2} \xi_2 \sqrt{B_2} d_{\xi_2} B_2 + \xi_2 d_{\xi_2} B_2 + 2(1+\nu_2) B_2 - \sigma B_2^{3/2} = 0\,. \tag{21}$$

the nontrivial solution of which must satisfy the same conditions as in the previous case:

$$d_{\xi_2} B_2(0) = d_{\xi_2} B_2^{\frac{3}{2}}(1) = B_2(1) = 0. \tag{22}$$

Figure 7(a-d, curves n = 2) shows the solution B_2 (ξ_2) for the cases σ = 1.25, 2.5, 5.0, 10.0.

Multiplying Equation (21) by ξ_2 and integrating it from 0 to 1 with a consideration for conditions (22), we find that at arbitrary positive values of parameter σ, ν_2 would be positive and $\nu_2(\sigma) \approx 0.0883\sigma - 0.000833\sigma^2$, while n_2 would be finite in relation to all σ and exists within the limits $1/2 < n_2 < 1$, (see Figure 6, n = 2). Moment M_2 remains constant for the cylindrical turbulent region under examination here, taking the form

$$M_2 = I_2 a^{2\nu_2}. \tag{23}$$

Expansion (16) is valid in relation to the solution of B_2 (ξ_2) in the vicinity of the boundary with the non-turbulent region ($\xi_2 \rightarrow 1$) that is, near the boundary the solution to the two-dimensional problem coincides with the solution to the one-dimensional problem with an accuracy of third-order terms containing the variable ((1 - ξ_2)/2). A contribution to subsequent terms of the expansion is made by the influence of the boundary's curvature and by dissipation. An analysis of dimensionalities leads to the following formula for the mean moment $\overline{M_2}$:

$$\overline{M_2} = \chi_2 \frac{q_s^{\frac{2}{3}(3+2\nu_2)}}{(\beta\delta T)^{2+2\nu_2}}. \tag{24}$$

where χ_2 is an empirical dimensionless constant which may be assessed in the same way that constants χ_1 has been assessed in the case of the 1-D turbulent layer.

3.3. Spherical Turbulent Region

In the case of a two dimensional system of internal waves propagating along the lower boundary of the upper layer $z = h$ in the plane $(x, 0, y)$ or, as a limit case, the horizontally isotropic internal waves field, the breaking zones are the finite turbulent regions of seawater (until they interact with each other). Each such region develops independently in response to the action of a turbulence charge acquired in the initial moment from local breaking of an internal wave. Using the same assumptions as before, we will approximate such a breaking zone by a non-stationary model of turbulence in a certain spherical turbulent region with its center located at depth $z = h$. Within the region, turbulence differs from zero and is centrally symmetrical, while out of this region it equals zero. The breaking zone is represented here by points within a region taking values values $0 \le R < R_0(t)$ when $z > h$, where $R = \sqrt{x^2 + y^2 + z_1^2}$, and $R = R_0(t)$ is the breaking zone's boundary with non-turbulent liquid. As before, in this case the turbulent energy flux and turbulent energy itself, k_w, are equal to zero at the boundary with the non-turbulent liquid, while in view of the symmetry approximation of the breaking zone, the turbulent energy flux is equal to zero and $k_w \neq 0$ at the boundary of contact with the upper layer when $z = h$. Assuming the turbulence scale to be , $\ell = \gamma_\ell R_0(t)$, we get the following turbulent energy budget equation for variable k_w written in variables R and t:

$$\partial_t k_w = (\gamma_1 \gamma_\ell R_0(t) / R^2) \partial_R R^2 \sqrt{k_w} \partial_R k_w - (\gamma_2 / \gamma_\ell) k_w^{3/2} / R_0(t) , \tag{25}$$

which is supplemented by the following initial condition: the initial distribution of turbulent energy exhibits central symmetry, and is equal to zero outside a certain spherical region $0 < R < a$, where a is the initial radius of the turbulent region. This condition takes the form

$$k_w(R, 0) = \frac{I_3}{4\pi a^2} u_3(\tau_3), \ \tau_3 = \frac{R}{a},$$
$$I_3 = 4\pi \int_0^a k_w(R, 0) R^2 dR, \ \int_0^1 u_3(\tau_3) \tau_3^2 d\tau_3 = 1, \tag{26}$$

where $u_3(\tau_3)$ is a dimensionless function of the argument τ_3, equal to zero when $\tau_3 \ge 1$, and I_3 is the total charge of turbulent energy in a spherical region with radius a, generated by the breaking of an internal wave. This situation is fully similar to the cases examined in the previous sub-sections. Assuming the existence of incomplete self-similarity, the asymptotic solution may be represented in the following form:

$$k_w(r,t) = \frac{4A_3^2 B_3(\xi_3)}{[(5+2\nu_3)\gamma_1\gamma_\ell]^2 t^{2n_3}},$$
$$\xi_3 = \frac{R}{R_0(t)}, \quad R_0(t) = A_3 t^{1-n_3}, \quad n_3 = \frac{3+2\nu_3}{5+2\nu_3}, \tag{27}$$

where A_3 is a constant and $\nu_3(\sigma)$ is the eigenvalue of this nonlinear problem. The equation for the function $B_3(\xi_3)$ is obtained from (25) with a consideration of the formulation (26) – (27) and it has the form

$$\xi_3^{-2} d_{\xi_3} \xi_3^2 \sqrt{B_3} d_{\xi_3} B_3 + \xi_3 d_{\xi_3} B_3 + (3+2\nu_3)B_3 - \sigma B_3^{3/2} = 0\,. \tag{28}$$

Its nontrivial solution must satisfy conditions similar to conditions (14) and (22):

$$d_{\xi_3} B_3(0) = d_{\xi_3} B_3^{\frac{3}{2}}(1) = B_3(1) = 0. \tag{29}$$

Figure 7(a-d, curves $n = 3$) shows the solution $B_3(\xi_3)$ for the cases $\sigma = 1.25$, 2.5, 5.0, 10.0.

It follows from Equation (28) and conditions (29) that when $\sigma > 0$, $\nu_3 > 0$ and n_3 is within the limits $3/5 < n_3 < 1$, and $\nu_3(\sigma) \approx 0.075\sigma$ (see Figure 6, $n = 3$). In the case of a spherical turbulent region, moment M_3 remains constant, taking the form

$$M_3 = I_3 a^{2\nu_3}. \tag{30}$$

The solution of $B_3(\xi_3)$ in the vicinity of the boundary with the non-turbulent region ($\xi_3 \to 1$) coincides with the solution of the 1D-turbulent problem with an accuracy up to third-order terms. The expansion (16) is also valid in relation to $B_3(\xi_3)$ with the indicated accuracy.

Dimensional analysis produces the following formula for the mean moment $\overline{M_3}$:

$$\overline{M_3} = \chi_3 \frac{q_s^{\frac{2}{3}(4+2\nu_3)}}{(\beta\delta T)^{3+2\nu_3}}, \tag{31}$$

where χ_3 is an empirical dimensionless constant for the three dimensional case.

Both types of localized zones of breaking internal waves examined in cases 3.2 and 3.3 may be encountered in the ocean. The turbulence charge intensities characterizing these breaking zones: I_2 and I_3, and their dimensions are random functions of time and spatial coordinates x, y in the general case (a dependence upon the vertical coordinate z comes into being when background stratification that is unperturbed by wind-caused mixing exists beneath the upper layer). Given a sufficiently dense distribution of such breaking zones at the lower boundary of the upper layer, with time they begin to interact with one another, in the final analysis forming a certain irregular turbulent interlayer at the lower boundary of the upper quasi-uniform layer, separating this upper layer from ocean water proper. Turbulence in this interlayer is continually fed by the action of breaking internal waves, which break at random moments and at random points in space. Such statistically steady-state situation may be described using the local breaking model, examined in case 3.1, if we assume that the initial charge I_1 created by a breaking internal wave and the initial dimension a are random functions of time and spatial coordinates. In this case, the knowledge of the statistical properties of variables I_1 and a makes it possible to completely determine the statistical properties of turbulence in the breaking zone, k_w, and the thickness of this zone, z_o. Unfortunately, the available experimental data on the breaking of internal waves are not yet sufficient to determine the statistical properties of variables I_1 and a, and for obtaining final conclusions on the properties of turbulence associated with isolated cases of breaking.

Figure 7 shows the examples of the non- dimensional TKE B_n profiles, ((a, b, c, d) - equations (13), $n = 1$; (21), $n = 2$; (28), $n = 3$; $\sigma = 1.25, 2.5, 5.0, 10$); (e) - the TKE profile B_2 and (f, $n = 2$) - the TKE k_w (0, t) of the 2D-region temporal evolution with the TKE data of shear – free turbulence experiment [Naudascher, 1965]. One can see that the TKE profiles in the normalized form in Figure 7(e) are not much sensitive to variations n, ν_2, σ, and well supported by the measurements. The data [Naudascher, 1965] correspond to $n_2 \approx 0.7$. Then there are $\nu_2 \approx 4/3$, $\sigma \approx 50/3$ that leads to the growth $r_0(t)$ of the breaking turbulent 2D-region as following to the power law "$r_0(t) \sim t^{0.3}$" and the decay of the TKE k_w (0, t) as following to the power law "$k_w(0, t) \sim t^{-1.4}$" (Figure 7,

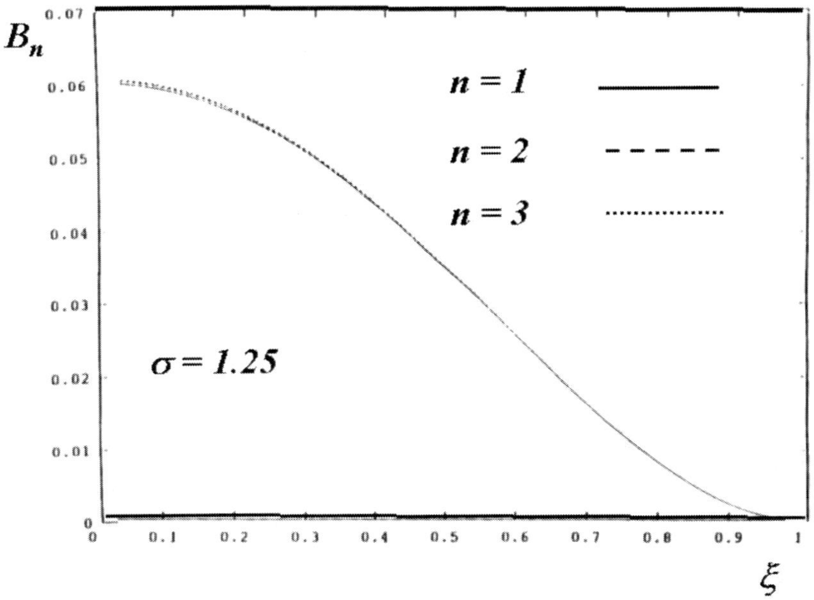

a

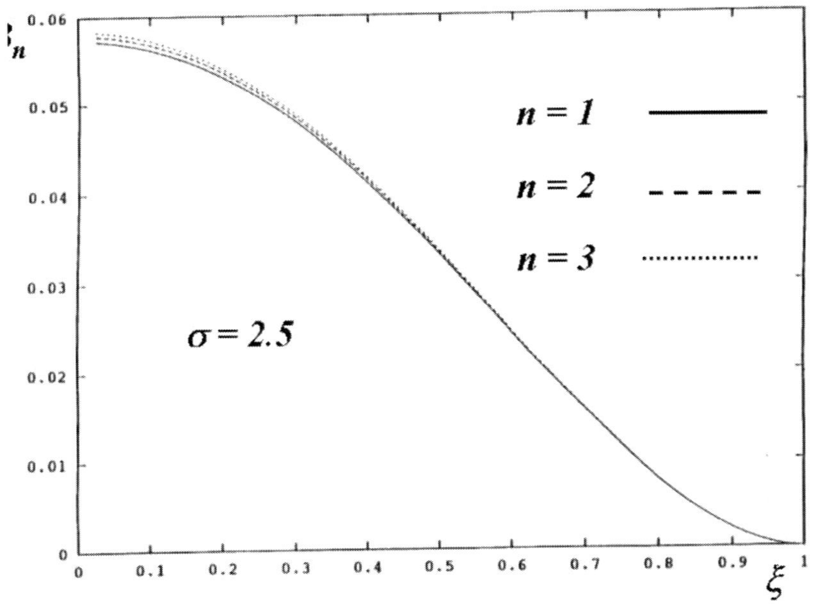

b

Figure 7. (Continued)

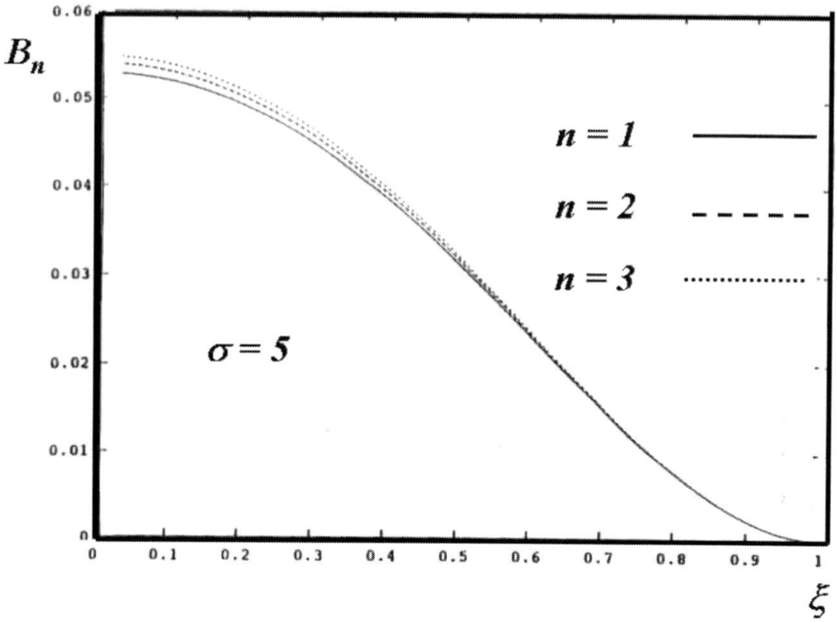

B_n
0.06
0.05
0.04
0.03
0.02
0.01
0
$n = 1$
$n = 2$
$n = 3$
$\sigma = 5$
0 0.1 0.2 0.3 0.4 0.5 0.6 0.7 0.8 0.9 1
ξ

c

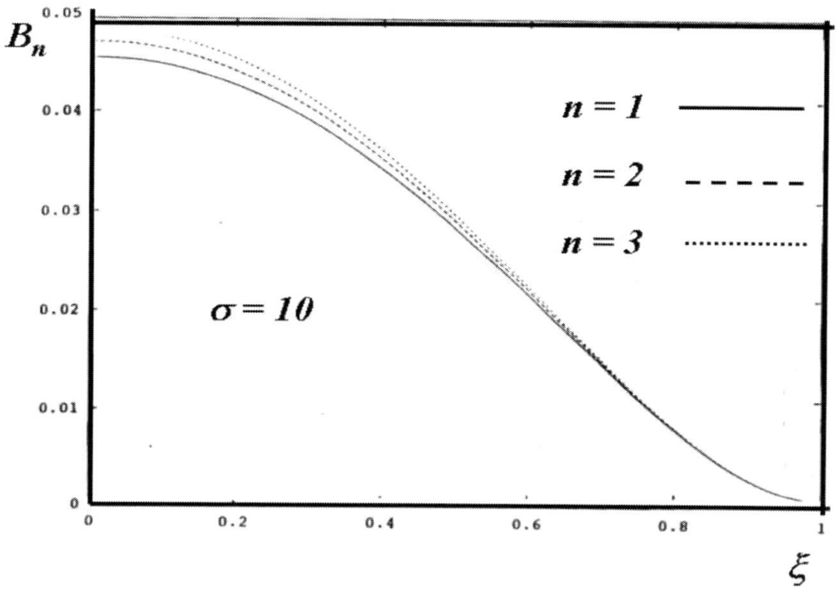

B_n
0.05
0.04
0.03
0.02
0.01
0
$n = 1$
$n = 2$
$n = 3$
$\sigma = 10$
0 0.2 0.4 0.6 0.8 1
ξ

d

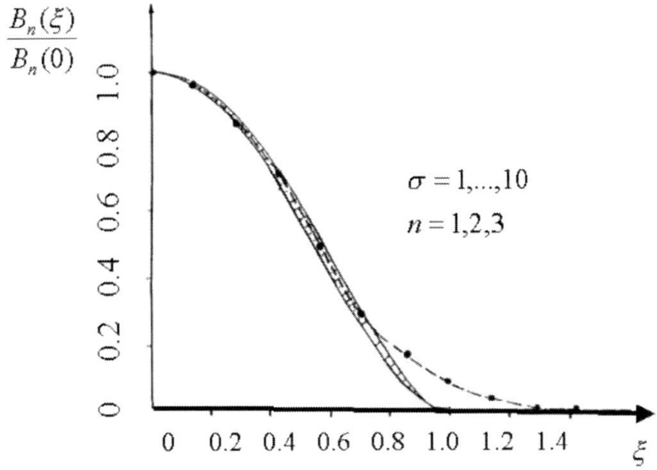

e

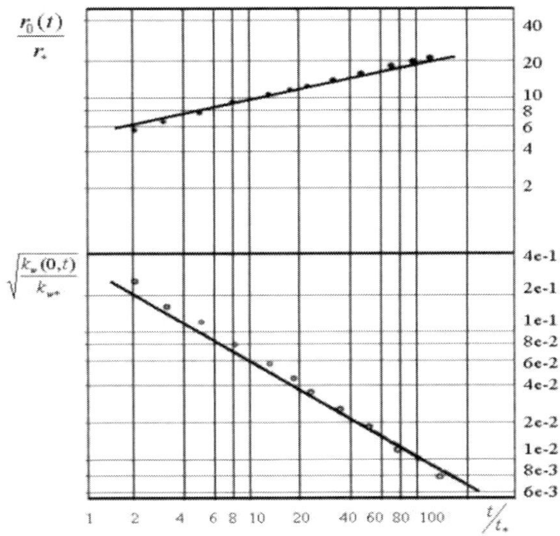

f

Figure 7. (a, b, c, d) - Examples of the non- dimensional TKE B_n profiles, (equations (13), $n = 1$; (21), $n = 2$; (28), $n = 3$; $\sigma = 1.25, 2.5, 5.0, 10$); (e) - profile of TKE B_2 and (f) - the TKE temporal evolution with the TKE data of shear – free turbulence experiment [Naudascher, E., 1965].

(e, f)). The value of the parameter $\sigma \approx 50/3$, which is consistent with the Naudascher's data, corresponds to the numerical value of the closure constant $\gamma_\ell \approx 0.073$ for typical constants $\gamma_1 = 1$, $\gamma_2 = 0.09$ [Hoffman, 1989] that means the efficient turbulent mixing scale is $\approx 0.073 r_0(t)$ in this experiment.

4. Interaction of the Upper Quasi-Uniform Layer with the Turbulence of a Breaking Zone

When a turbulent region arises at the lower boundary of the upper layer and then spreads out of the limit of this layer, its dynamic and thermodynamic equilibrium becomes locally disturbed (at a scale commensurate with the scale of the turbulent region formed by a breaking wave). This disturbance is elicited by the exchange of heat and energy between the upper layer and the underlying liquid within the limits of the turbulent region made by a breaking internal wave. The balance of heat and energy arising in this case may be represented approximately as

$$q_s - q_h = \kappa_2 g \alpha (T_s - T_\infty) h \sqrt{k_s}\,, \tag{32}$$

$$G_h = \rho_a c_{pa} C_T (T_a - T_s)\,, \tag{33}$$

where q_h and G_h are the turbulent fluxes of turbulent energy and heat at depth $z = h$ the boundary of contact between the upper layer and the turbulent zone made by the breaking waves. In the case of cylindrical turbulent zones the contact surface is a band with width $D_2 = 2r_0$ (Section 3.2), oriented perpendicular to the direction of y. In the case of spherical turbulent zones (Section 3.3), these are circles with diameter $D_3 = 2R_0$. In the case of the statistically steady and horizontally uniform occurrences of the internal wave breaking when the breaking turbulent regions overlap, the lower boundary of the upper layer transforms into the breaking turbulence layer (Section 3.1). Wherever breaking is absent, the fluxes q_h and G_h equal to zero. Equation (32) is a generalization of the balance Equation (5) for the case $q_h \neq 0$, and it has the same physical meaning. In this case, however, k_s no longer equals k_e due to the interaction of the upper layer with the breaking zone. According to (33), the heat flux G_h at the breaking zone is expended in increasing the temperature of the external cold seawater that is captured by the breaking turbulence and disturbs the heat balance in the region of interaction and then $T_s < T_a$ and the heat flux from the atmosphere (the left side of (33)) compensates these losses.

The simplest parameterization of the fluxes energy q_h and heat G_h at the breaking zone is

$$q_h = \kappa_3 \sqrt{k_w k_s} (\sqrt{k_s} - \sqrt{k_w}), \tag{34}$$

$$G_h = \kappa_4 \rho_w c_{pw} \sqrt[4]{k_w k_s} (T_s - T_w), \tag{35}$$

where κ_3 and κ_4 are dimensionless empirical coefficients, ρ_w and c_{pw} are the density of water and its specific heat, respectively, k_s is the turbulent kinetic energy of the upper layer at the breaking zone and k_w is the turbulent kinetic energy of the breaking zone averaged over the entire thickness of the breaking zone. In our case the seawater temperature of breaking zone $T_w = T_\infty$. The parameterization (34)-(35) provides a correct physical meaning of the quantitative representation of the fluxes energy q_h and heat G_h, in the sense of their direction and magnitude, which are controlled by the differences in the right hand sides of (34)-(35) and the factor k_w that reduces these fluxes to zero in the case of absence of the breaking turbulence ($k_w = 0$).

Using Equations (32) - (35), the dependences of k_s and T_s on the turbulent kinetic energy k_w of the breaking zone are as follows

$$v_s = \frac{1}{2}\left[\sqrt{\left(\frac{v_w^2 - \theta\kappa_1^{-3/2}\kappa_3^{-1}}{v_w}\right)^2 + \frac{4}{\kappa_1^{3/2}\kappa_3^{1}v_w}} + \frac{v_w^2 - \theta\kappa_1^{-3/2}\kappa_3^{-1}}{v_w}\right], \tag{36}$$

$$\theta = \frac{a_1}{a_1 + \sqrt{v_w v_s}}, a_1 = \frac{b_1 U_a}{\kappa_1^{1/2} q_s^{1/3}}, \tag{37}$$

$$v_s = \sqrt{\frac{k_s}{k_e}}, v_w = \sqrt{\frac{k_w}{k_e}}, \theta = \frac{T_s - T_\infty}{T_a - T_\infty}, b_1 = \frac{\rho_a c_{pa} C_T}{\rho_w c_{pw} \kappa_4}. \tag{38}$$

Formula (36) shows that when $0 \le v_w \le 1$ and θ is close to unity ($v_w << 1$, or because $a_1 >> 1$), v_s decreases from $vs = 1$ to some minimum value with growth v_w, and then it increases once again rising to $v_s = 1$when v_w is close to unity (if $\theta = 1$ and $v_w = 1$ then $v_s = 1$, since in this case k_w does not influence on stratification in (32)). At large values v_w, the disturbed TKE v_s is asymptotically proportional to v_w ($v_s \approx v_w$ when $v_w \to \infty$). Thus, in the latter limiting case, the effect of surface waves breaking is insignificant and the breaking of internal waves defines entirely the turbulent kinetic energy k_s. Under natural conditions this limiting situation does not look realistic - the most natural cases corresponds to $v_w \leq 1$.

The temperature of the upper layer, consequently the temperature of the ocean surface above the breaking zone, decreases monotonously with increase v_w. Because the turbulent breaking zone enlarges with time and the turbulence within it weakens, the anomalies in the upper layer's turbulence field, characterized by the variable v_s, and anomalies in the temperature field characterized by the variable θ will decrease in magnitude, although the spatial dimension of these non-uniformities will grow.

According to [Fedorov, 1981], the non-uniformities of surface temperature recorded in therange of the internal waves scales are about 0.1°C. Taking this value as a characteristic magnitude of temperature perturbation, we come up to the estimate that the parameter $a_1 = b_1 U_a / \kappa_1^{1/2} q_s^{1/3}$ takes order of magnitude of $a_1 \sim 10^2$, and that corresponds to the expected real values of $v_w \sim 1$, $\theta \sim 1$. For this reason, the variable v_s would be of an order or less than unity in situations realistically encountered in the ocean.

The relationships presented in Section 3 and the equations (36)-(38) solve the problem posed in this paper. In particular, it follows from these formulas and from the relationships of Section 3.2, that in the case of a breaking zone having the form of a long cylinder, slicks-areas of smoothing of surface waves with the wave lengths of several centimeters long assume the form of alternating bands perpendicular to wave propagation. In the case where the scales of the breaking zone are approximately the same in all directions in a top view, the breaking zone may be approximated by a hemisphere. The relationships of Section 3.3 and formulas (36)-(38) show that in this case slicks have a ring or circular form. In the limit case of random wave breaking events with differently shaped zones of breaking turbulence, the form of the slicks is expected as a set of random plane geometrical objects.

The numerical calculations of the surface disturbances of both the disturbed turbulent kinetic energy v_s (Figure 8) and the temperature θ of the upper layer (Figure 9) as functions of the intensity of turbulent kinetic energy inside the internal wave breaking zone v_s have been carried out using (36)-(38)

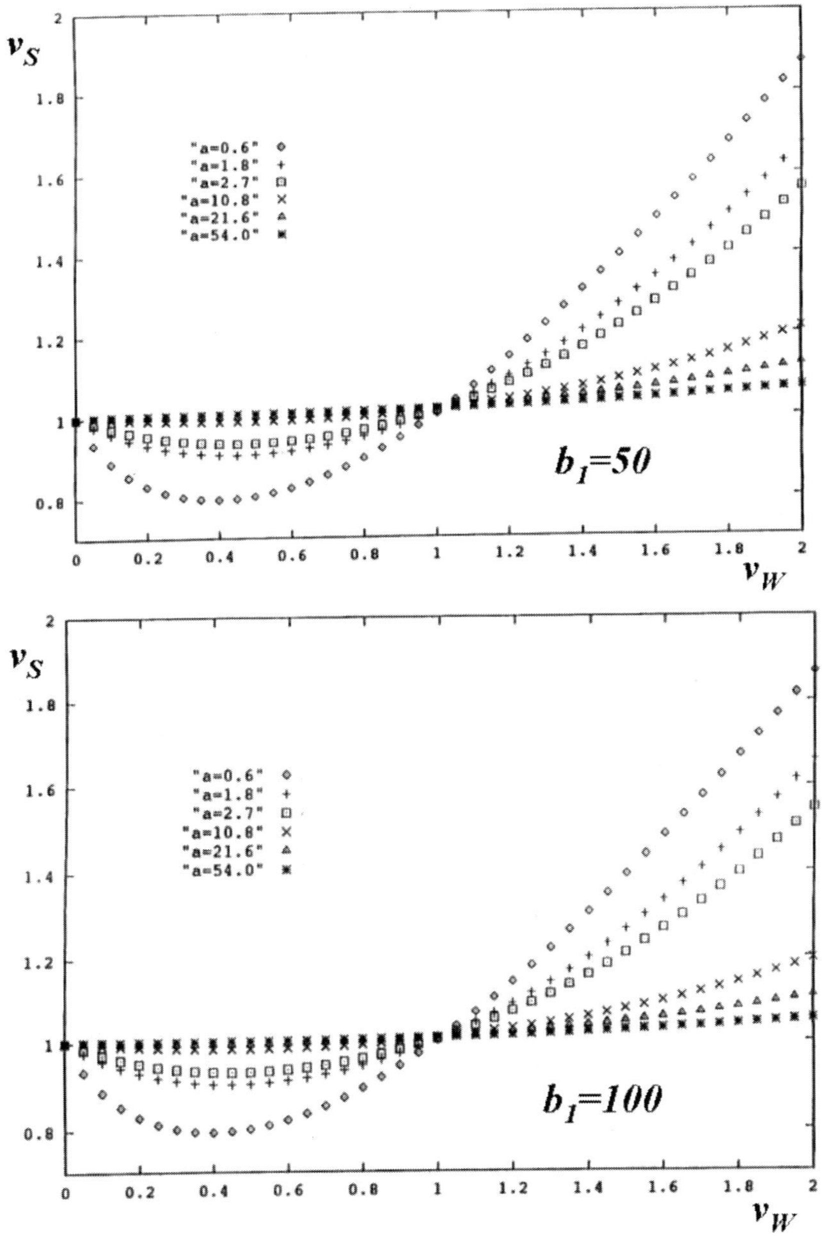

Figure 8. The example of calculation the disturbed TKE v_s versus of the internal breaking TKE v_w equations ((36) – (38)) for different numerical values of the parameters parameters $a = \kappa_1^{-3/2}\,\kappa_3^{-1}$ and $b_1 = \rho_a c_{pa} C_T \,/\, \rho_w c_{pw} \kappa_4$.

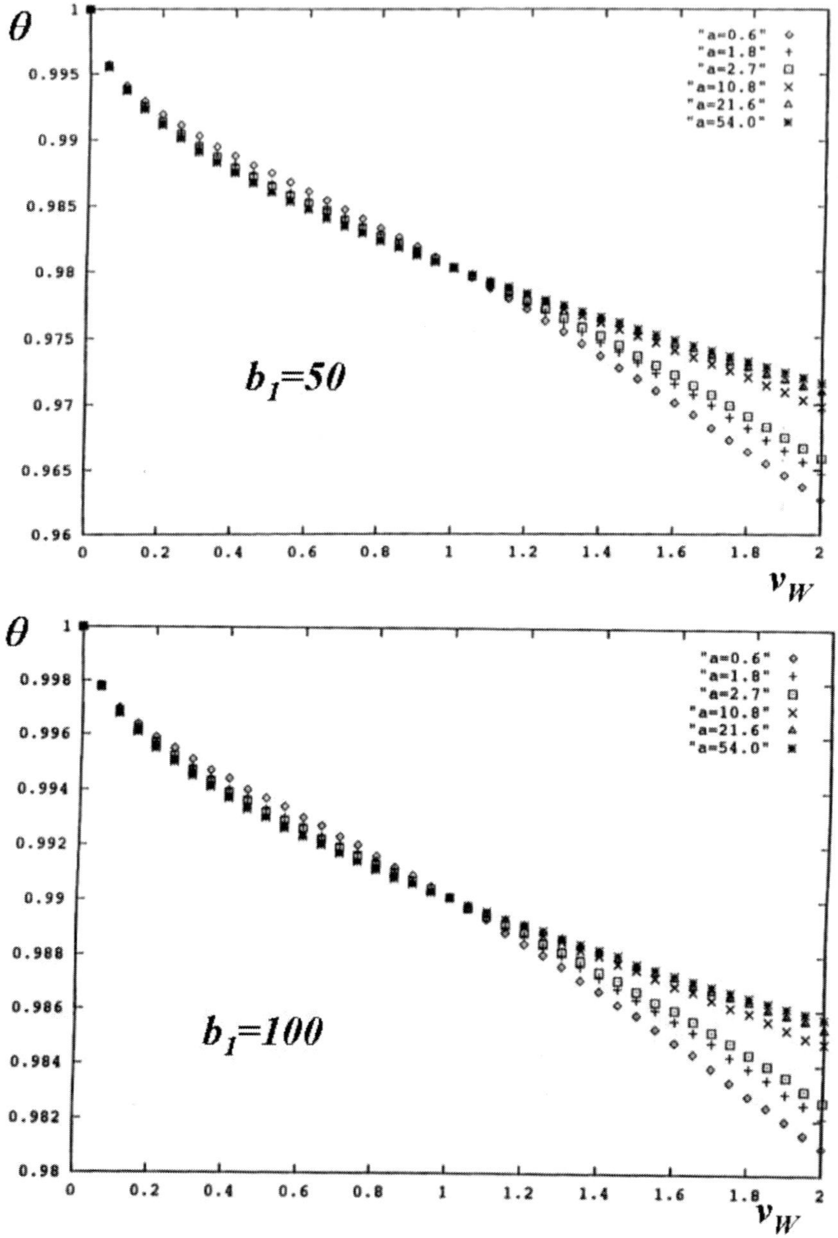

Figure 9. The example of calculation the disturbed temperature of the upper layer θ versus of the internal breaking TKE v_w (equations (36)–(38)) for different numerical values of the parameters $a = \kappa_1^{-3/2}\,\kappa_3^{-1}$ and $b_1 = \rho_a c_{pa} C_T \,/\, \rho_w c_{pw} \kappa_4$.

The dependences in Figure 9, which is calculated from equation (38), describe not only the surface temperature disturbances but also the surface heterogeneities of the biogenic salts (or mineral nutrition, Section 5).

The region of surface smoothness is well seen in Figure 8 for values of the quantity v_w such that $v_w < 1$. There is a minimum v_s which decreases from $v_{s\min} \le 1$ to $v_{s\min} \approx 0.8$ simultaneously with the decrease of the parameter of interaction of the upper layer turbulence and the turbulence of the internal wave breaking zones $a = \kappa_1^{-3/2}\kappa_3^{-1}$ from $a = 54$ to $a = 0.6$. The parameter a for real situations has small values in that the disturbance value can reach 20% of the ambient value. The temperature disturbance of this example is not more than 1% from the undisturbed temperature difference $T_a - T_\infty$ (Figure 9). Therefore the change of the parameter b_1 from $b_1 = 50$ to $b_1 = 100$ does not affect on v_s (Figure 8). In the case of biogenic salts (mineral nutrition), the magnitude of disturbance takes order of about 1% of the biogenic salts concentration in the main body of the ocean water under the mixing layer. The presence of biogenic substances at the ocean surface is the evidence of the existence of the "living" internal wave breaking zones. For instance in the observations [Bock et. al, 1990; Moum et. al, 1990; Wijesekera and Willon, 1990] it was found that the well-stratified pycnocline region was populated by intermittent, highly energetic, turbulent bursts with lifetimes of 2-3 hours, separated in time by several hours. During periods of moderate (or greater) winds, high frequency, high amplitude, internal gravity waves were observed, primarily at night. Regions of intense wave activity were generally mixing above background levels. Observations indicated that the internal waves increase turbulence levels in the pycnocline. These results demonstrate a strong correlation between local surface smoothness and surface enrichment of biological and chemical materials, and coincidence between chemical enrichment and underlying temperature discontinuities, suggest that subsurface physical processes influence slick formation, including transfer of chemical components to the ocean surface.

5. The Horizontal Diffusion Effect

Horizontal diffusion always occurs in the upper layer, increases the scales of horizontal non-uniformities and decreases their level. The random field of the currents of the upper layer causes the horizontal diffusion. As it is shown in [Monin and Ozmidov, 1985; Ostrovskiy and Piterbarg, 1985], the approximation of horizontal diffusion by a horizontal diffusion coefficient takes into account the random nature of wind currents as the mesoscale horizontal turbulence of the upper layer.

We shall now consider the formation of horizontal non-uniformities of temperature and biogenic salts in the same model approach of the upper layer as in Sections 2-4, but with the developed horizontal turbulence characterized by the horizontal diffusion coefficient D. The upper mixing layer has a thickness h, which from equations (5)-(7) is:

$$h \approx 3 \times 10^{-3} \frac{U_a^2}{g\alpha(T_a - T_\infty)}, \tag{39}$$

where U_a is wind velocity at a standard measuring height $z = z_a$, g is the acceleration due to gravity, α is the coefficient of thermal expansion, T_a is air temperature at a height z_a, and T_∞ is the temperature of the ocean water of the main ocean interior below the upper layer. The biogenic salts, which we shall characterize by the concentration C, are generally placed under the upper mixing layer, where we shall assume they have a constant concentration C_∞. The transfer of both heat and biogenic salts across the density jump layer occurs through the turbulent regions that are created by the breaking internal waves – or the heat-mass transfer occurs through the turbulent holes in the pycnocline. In the upper layer, the biogenic salts are utilized by phytoplankton in the photosynthesis reaction. We assume that the photosynthesis reaction layer coincides with the upper mixing layer. The fluctuations of phytoplankton concentration are much less than its mean value. The biogenic salts restoration occurs in the ocean interior, and the radiation energy flux is characterized by some effective constant value. These assumptions are the rough approximation of the real ocean situation [Seomina, 1977; Ivanenkov, 1979; Benilov, 1989; Hood et al., 1990; Dickey, 1991]. We shall denote the biogenic salts concentration at the upper layer and the biogenic utilization time by C_s and τ_c, respectively. Thereupon the equations for the description of the variability of the upper layer temperature T_s and biogenic salts concentration C_s are

$$\partial_t T_s = D\Delta T_s - \partial_z G / \rho_w c_{pw}, \tag{40}$$

$$\partial_t C_s = D\Delta C_s - \partial_z Q_C - \tau_C^{-1} C_s , \tag{41}$$

where $\Delta = \partial^2_{xx} + \partial^2_{yy}$ is the two-dimensional Laplace operator, and G and Q_C are the vertical turbulent fluxes of heat and biogenic salts.

The upper mixing layer is unbounded in the horizontal planes and the fluxes G and Q_C do not depend on the horizontal coordinates (or they are statistically steady and spatially uniform functions of the horizontal coordinates). On the ocean surface the condition (3) for heat flux G is valid, and the obvious condition for the biogenic salts flux Q_C, that is

$$Q_C\big|_{z=0} = 0 \tag{42}$$

At the lower boundary of the upper layer the heat flux $G(z = h) = G_h$ is given by the equation (35), and the similar assumptions can be applied for the turbulent flux of the biogenic salts $Q_C\,(z = h) = Q_{Ch}$ that takes the form

$$Q_C\big|_{z=h} = Q_{Ch} = -\kappa_4 \sqrt[4]{k_w k_s}\,(C_\infty - C_s)\,. \tag{43}$$

Following to the physical statement of the problem discussed in Sections 2-4, we shall assume that the temperature T_s and concentration C_s are functions of the time t and horizontal coordinates x and y. The turbulent fluxes G and Q_C must be linear functions of the vertical coordinate z to satisfy these requirements. Taking into account the equations (3), (35), (42), (43) we find that G and Q_C in the upper layer take the form

$$G = G_s + (G_h - G_s) z / h\,, \tag{44}$$

$$Q_C = Q_{Ch} z / h\,. \tag{45}$$

Substitution (44) and (45) in equations (40)-(41) and taking into account (3), (35), (42), (43) reduce the equations (40) and (41) to the following form:

$$\begin{aligned} \partial_t T_s &= D\Delta T_s - \left[C_T \frac{\rho_a c_{pa} U_a}{\rho_w c_{pw} h} + \kappa_4 \frac{\sqrt[4]{k_w k_s}}{h} \right] T_s \\ &+ \left[C_T \frac{\rho_a c_{pa} U_a}{\rho_w c_{pw} h} T_a + \kappa_4 \frac{\sqrt[4]{k_w k_s}}{h} T_\infty \right] \end{aligned}, \tag{46}$$

$$\partial_t C_s = D\Delta C_s - \left[\frac{1}{\tau_C} + \kappa_4 \frac{\sqrt[4]{k_w k_s}}{h}\right] C_s + \kappa_4 \frac{\sqrt[4]{k_w k_s}}{h} C_\infty . \tag{47}$$

These equations coupled with the energy budget equation (32) and results of Section 3 describe the variability of T_s and C_s caused by the turbulent kinetic energy bursts k_w of internal wave breakings.

The use the new notations of the temperature and biogenic salts fields in the form

$$\theta = \begin{pmatrix} \dfrac{T_s - T_\infty}{T_a - T_\infty}, \\ \dfrac{C_\infty - C_s}{C_\infty} \end{pmatrix}, \tag{48}$$

$$\tau_\theta = \begin{pmatrix} \dfrac{\rho_w c_{pw} h}{C_T \rho_a c_{pa} U_a}, for T_s, \\ \tau_C, \quad for C_s \end{pmatrix}, \tag{49}$$

$$w = \kappa_4 \sqrt[4]{k_w k_s} , \tag{50}$$

reduces two equations (46) and (47) for T_s and C_s to one equation for the variable θ, which is

$$\partial_t \theta = D\Delta\theta - \frac{\theta - 1}{\tau_\theta} + \frac{w}{h}\theta . \tag{51}$$

Similar equations are applied to sea-surface temperature anomalies caused by the thermocline variability [Frankingnoul and Hasselmann, 1977; Ostrovskiy and Piterbarg, 1985]. In the present problem, the variability of θ is caused by the internal waves breaking that is characterized by the function w, which is a random function in the real oceanic situations. The steady and horizontally homogeneous solution θ_0 of the equation (51) is

$$\theta = \theta_0 = \frac{1}{1+\dfrac{w\tau_\theta}{h}} = \begin{pmatrix} \dfrac{1}{1+\sqrt[4]{k_w k_s} / a_1 U_a}, & for T_s \\ \dfrac{1_s}{1+\kappa_4 \tau_C \sqrt[4]{k_w k_s} / h}, & for C_s \end{pmatrix}. \quad (52)$$

This solution coincides with the solution (37) for T_S. In the case of the concentration of the biogenic salts C_s, the parameter $h/\kappa_4\tau_C$ replaces the parameter a_1U_a of the temperature formula. The variations of T_s and C_s as the response on the turbulent energy burst k_w are different due to the numerical value of the parameter τ_θ / h , which are different for the temperature field T_s and the biogenic salts C_s.

Let us consider now the case of a random field *w*, which is a model approximation of turbulent bursts caused by random events of the internal waves breaking. The random field *w* is assumed as a statistically stationary and spatially homogeneous random field. The turbulent regions of the pycnocline occur intermittently and are characterized by the intermittence coefficient, *p*, which is the mean fraction of the turbulent area caused by the internal wave breaking. The coefficient *p* takes the order ~ 10^{-2} in the ocean pycnocline [Grant, Moilliet and Vogel, 1968; Marmorino, 1990]. The random field of internal wave breaking turbulence *w* causes the variability of the upper layer fields θ, which are defined by equations (48)-(51).

The mean value of *w* at the turbulent area is $\overline{w}$, and the total mean of *w* over the entire pycnocline area populated by the events of the internal waves breaking is $< w >$. The connection between $< w >$ and $\overline{w}$ is given by the obvious equality

$$< w > = p\overline{w}, \quad (53)$$

and then the fields *w* and θ take the form

$$w = < w > + w', \ \theta = < \theta > + \theta', \quad (54)$$

where $< \theta >$ is the mean value of θ-field, w' and θ' are their fluctuations with zero mean values, and w' is in the range $-p\overline{w} \le w' \le (1-p)\overline{w}$. Then, using also the results of Section 4, the following estimates are acceptable

$$|w'| \le (1-p)\overline{w} \approx \overline{w}, \ < \theta > \sim 1, \ |\theta'| << 1, \quad (55)$$

where upper estimate $|w'| \sim \overline{w}$ can be observed as a burst of turbulence in the breaking zone. Keeping this fact in mind the correlation $< w'\theta' >$ can be estimated by the order of its magnitude as $< w'\theta' > \sim p\overline{w}|\theta'| << p\overline{w} < \theta >$. Then the substitution (53) into the equation (51) and the use the estimates (55) allow finding the mean value $< \theta >$ and the equation for fluctuation θ' which are, correspondingly, as

$$< \theta > \approx \frac{1}{1 + \frac{< w > \tau_\theta}{h}} = \frac{1}{1 + p\frac{\overline{w}\tau_\theta}{h}}, \tag{56}$$

$$\partial_t \theta' = D\Delta\theta' - \frac{\theta'}{\tau_\theta} + \frac{w'}{h} < \theta > . \tag{57}$$

Comparing (56) with (52) one can see that the mean value $< \theta >$ and the steady solution θ_0 are alike, but expression (52) is the exact steady solution while the expression (56) is an asymptotic approach under conditions (55).

One can see (equations (56)-(57)) that if there is strong intermittence ($p \to 0$) then the upper mixing layer will be impermeable to interchange of the heat and the biogenic salts with the ocean interior, $< \theta > \to 1$ and $(w', \theta') \to 0$. The last asymptotic means that initial conditions of the equation (57) and very rare bursts of internal wave breaking turbulence degenerate at the time scales less than the time intervals between the turbulent bursts. The conditions (55) are well fulfilled for realistic numerical values of parameters of the present problem. Therefore (56) can be taken as an acceptable estimate of the mean value $< \theta >$ caused by the random field of internal wave breaking turbulence and (57) describes the variability of θ-field.

Equation (57) allows finding the statistical connections of the characteristics of fluctuations w' and θ'. In accordance with the theory of random fields [Monin and Yaglom, 1987] the random fields $w'$and θ' can be taken in the form of an extension of Fourier integrals on random fields, the Fourier-Stieltjes integrals:

$$w'(\boldsymbol{r}, t) = \int e^{i(\boldsymbol{kr} - \omega t)} dZ_w(\boldsymbol{k}, \omega), \quad \theta'(\boldsymbol{r}, t) = \int e^{i(\boldsymbol{kr} - \omega t)} dZ_\theta(\boldsymbol{k}, \omega), \tag{58}$$

where dZ_w ($\boldsymbol{k}$, ω) and dZ_θ ($\boldsymbol{k}$,ω) are the Fourier-Stieltjes measures of the random fields w' and θ' (they have meaning of random amplitudes of the fields w' and θ' corresponding to spatial-temporal scales $\boldsymbol{k}$ and ω), $\boldsymbol{r} = \{r_x, r_y\}$ is a horizontal vector, $\boldsymbol{k} = \{k_x, k_y\}$ is a wave vector, ω is a frequency, and the integrals are evaluated over the entire wave-number vector space and the frequency space. The Fourier-Stieltjes measures dZ_w ($\boldsymbol{k}$, ω) and dZ_θ ($\boldsymbol{k}$,ω) are random complex functions of their real arguments $\boldsymbol{k}$ and ω, and their complex conjugative functions are $dZ^*_w(\boldsymbol{k},\omega)$ and $dZ^*_\theta(\boldsymbol{k},\omega)$. Then the spectra $\Phi_w(\boldsymbol{k},\omega)$, $\Phi_\theta(\boldsymbol{k},\omega)$ and variances $\sigma_w^2 = < w'^2 >$, $\sigma_\theta^2 = < \theta'^2 >$ are

$$< dZ_q(\boldsymbol{k},\omega) dZ_q^*(\boldsymbol{k},\omega) > = \Phi_q(\boldsymbol{k},\omega) d\boldsymbol{k} d\omega , \tag{59}$$

$$\sigma_q^2 = \iiint \Phi_q(\boldsymbol{k},\omega) d\boldsymbol{k} d\omega , \qquad d\boldsymbol{k} = dk_x dk_y , \qquad q = (w,\theta), \tag{60}$$

where the integrals are evaluated over the entire wave-number vector space and the frequency space.

Substituting (58) into equation (57) and taking into account the definition (59) we find that the spectrum $\Phi_\theta(\boldsymbol{k},\omega)$ is

$$\Phi_\theta(\boldsymbol{k},\omega) = \frac{< \theta >^2}{h^2[(k^2 D + \tau_\theta^{-1}) + \omega^2]} \Phi_w(\boldsymbol{k},\omega), \; k = \sqrt{k_x^2 + k_y^2} . \tag{61}$$

Then it follows that the spatial inhomogenities with scales $k^{-1} >> \sqrt{D\tau_\theta}$ are not subjected to the diffusion influence. The horizontal diffusion smoothes out heterogeneities of θ'. The degeneration effect with the characteristic time τ_θ is essential for spatial-temporal scales $k^{-1} >> \sqrt{D\tau_\theta}$ and $\omega^{-1} >> \tau_\theta$.

The variance σ^2_θ and the one-dimensional spectrum E_θ (k_x) can be calculated using the expression (61). The 1D- spectrum $E_\theta(k_x)$ is

$$E_\theta(k_x) = \int_{-\infty}^{\infty}\int_{-\infty}^{\infty} \Phi_q(k_x, k_y, \omega) d\omega dk_y \,, \tag{62}$$

where the inner integral

$$\int_{-\infty}^{\infty} \Phi_q(k_x, k_y, \omega) d\omega = F_\theta(k_x, k_y) = \frac{<\theta>^2}{h^2} \int_{-\infty}^{\infty} \frac{\Phi_w(\boldsymbol{k}, \omega) d\omega}{(k^2 D + \tau_\theta^{-1}) + \omega^2} \tag{63}$$

is the two-dimensional spatial spectrum $F_\theta(k_x, k_y)$ of the field θ'.

The quantities σ_θ^2 and $E_\theta(k_x)$ are most accessible statistical characteristics in field measurements. Their explicit formulas can be found considering some model approximations of the spectrum $\Phi_w(\boldsymbol{k}, \omega)$.

a. In the case when the spectral density of the field w' is concentrated at the wave number $\boldsymbol{k}_0 = \{k_{0x}, k_{0y}\}$ and frequency ω_0, and the spectrum $\Phi_w(\boldsymbol{k}, \omega)$ can be represented in the following form

$$\Phi_w(\boldsymbol{k}, \omega) = \frac{\sigma_w^2}{8}[\delta(\boldsymbol{k} - \boldsymbol{k}_0) + \delta(\boldsymbol{k} + \boldsymbol{k}_0)][\delta(\omega - \omega_0) + \delta(\omega + \omega_0)],$$
$$\delta(\boldsymbol{k}) = \delta(k_x)\delta(k_y), \tag{64}$$

where $\delta(x)$ is the Dirac δ -function, we find from (60)-(64) that the variance σ_θ^2 is

$$\sigma_\theta = \frac{<\theta> \sigma_w \tau_\theta}{h\sqrt{(k_0^2 D \tau_\theta + 1)^2 + (\omega_0 \tau_\theta)^2}}, \tag{65}$$

where σ_w^2 is the variance of the random field w, induced by the internal waves breaking. If the random field w can be characterized by the mean value $\overline{w}$ and the variance $\overline{w'^2}$ in the internal waves breaking zones, then the quantity σ_w^2 can be written in the form

$$\sigma_w^2 = p\overline{w'^2} + p(1-p)\overline{w}^2 . \qquad (66)$$

The equations (55), (65), (66) show that the surface fluctuations of the field θ cease to depend on the degeneration time τ_θ for the scales $\ell_0 = k_0^{-1}$, which are much less than the diffusion scale $L_{\theta D} = \sqrt{D\tau_\theta}$. The non-uniformities of θ with such scales are subjected to the influence of the diffusion coefficient D. The implicit dependence on τ_θ is from $< \theta >$. The increasing of intermittence ($p \to 0$) results in decreasing σ_w^2 and σ_θ^2. In the case of the irregular turbulent interlayer at the lower boundary of the upper layer ($p \to 1$; see Section 3), the value of σ_θ^2 is defined by the value of σ_w^2. In the region of large-scale $\ell_0 >> L_{\theta D}$ and small values of the frequency ω_0 ($\omega_0 << \tau_\theta^{-1}$), the value of σ_θ^2 increases due to increase of τ_θ, and does not depend on the diffusion coefficient D.

b. Let us assume that the random field w' is delta-correlated with the correlation function

$$B_{w,w}(\boldsymbol{r},t) = < w'(\boldsymbol{x}+\boldsymbol{r},t+\tau)w'(\boldsymbol{x},t) > = \sigma_w^2 \ell_w^2 \tau_w \delta(\boldsymbol{r})\delta(\tau), \qquad (67)$$

$$\boldsymbol{x} = \{x, y\}\ ,\ \boldsymbol{r} = \{r_x, r_y\},\ \delta(\boldsymbol{r}) = \delta(r_x)\delta(r_y),$$

where ℓ_w and τ_w are the correlation scales and $|\ \boldsymbol{r}\ | >> \ell_w$, $|\ \boldsymbol{\tau}\ | >> \tau_w$. The spectral density $\Phi_w(\boldsymbol{k},\omega)$ in this case takes the simple form

$$\Phi_w(\boldsymbol{k},\omega) = \frac{\sigma_w^2 \ell_w^2 \tau_w}{(2\pi)^3}, \qquad (68)$$

so that the one-dimensional spectrum $E_\theta(k_x)$ (Figure 10, $E_\theta(k_x) / E_\theta(0)$) is:

$$E_\theta(k_x) = \frac{E_\theta(0)}{\sqrt{k_x^2 D\tau_\theta + 1}},\ E_\theta(0) = \left(\frac{< \theta > \sigma_w \ell_w}{2h}\right)^2 \frac{\tau_w \tau_\theta}{2\pi\sqrt{D\tau_\theta}}. \qquad (69)$$

The formula for 1D-spectrum in (69) shows that if the turbulence of different regions is statistically independent, and the spatial and temporal scales of the field θ are essentially greater than the characteristic scales of the turbulent regions and their life-time (in the present example they are ℓ_w and τ_w) then there are two different regions of scales. In the long-wave region, where the wave number k_x is in the range $k_x << (D\,\tau_\theta)^{-1/2}$, the spectral density $E_\theta\,(k_x)$ does not depend on the wave number k_x and is proportional to the quantity $(\tau_\theta / D)^{1/2}$. In the second region, where the wave number k_x is in the range $(D\,\tau_\theta)^{-1/2} << k_x << \ell_w^{-1}$, the spectrum $E_\theta\,(k_x)$ follows the law "k_x^{-1}" and does not depend on τ_θ. The latter region is the region of diffusion action on the non-uniformities of the upper layer. Assuming that k_x has the maximum value $k_{x1,\max}$, which for simplicity can be taken as $k_{x1,\max} \approx \ell_w^{-1}$, then the variance σ_θ^2 takes the form

$$
\begin{aligned}
\sigma_\theta^2 &\approx <\theta>^2 \frac{\sigma_w^2 \tau_w}{4\pi D} \frac{\ell_w^2}{h^2} \ln\left(\frac{\sqrt{D\tau_\theta}}{\ell_w} + \sqrt{\frac{D\tau_\theta}{\ell_w^2} + 1} \right) \approx \\
&\approx <\theta>^2 \frac{\sigma_w^2 \tau_w}{4\pi D} \frac{\ell_w^2}{h^2} \ln\left(\frac{2\sqrt{D\tau_\theta}}{\ell_w} \right)
\end{aligned}
\tag{70}
$$

The formula (70) for the variance σ_θ^2 shows that in the case when the field w is the white noise, the variance, σ_θ^2 weakly depends on the time degeneration τ_θ and is influenced by the diffusion coefficient D. Equations (53), (56), (66), (69) and (70) define the basic characteristics of variability of the field θ through the statistical characteristics of the white noise w.

c. This case considers the combination of examples (a) and (b) in the form of spatial white noise which is concentrated at the frequency ω_0. The spectral function $\Phi_w(\boldsymbol{k},\omega)$ can be represented by the formula

$$
\Phi_w(\boldsymbol{k},\omega) = \frac{\sigma_w^2 \ell_w^2}{(2\pi)^2} \frac{1}{2} \left[\delta(\omega - \omega_0) + \delta(\omega + \omega_0) \right], \tag{71}
$$

then the two dimensional spectrum $F_\theta(\boldsymbol{k})$ defined by the equation (63) is

$$F_\theta(\boldsymbol{k}) = <\theta>^2 \frac{\sigma_w^2 \tau_\theta^2 \ell_w^2}{h^2 (2\pi)^2} \frac{1}{(k_x^2 D\tau_\theta + 1)^2 + (\omega_0 \tau_\theta)^2}, \qquad (72)$$

and then, using equation (61) and the definition of the one-dimensional spectra (62), we find that the spectral function $E_\theta(k_x)$ for the present case (Figure 10, $G_\theta(x)$) is

$$E_\theta(k_x) = <\theta>^2 \frac{\sigma_w^2 \tau_\theta^2 \ell_w^2}{h^2 (2\pi)^2} I_\theta(k_x, \tau_\theta, \omega_0), \qquad (73)$$

$$\begin{aligned} I_\theta(k_x, \tau_\theta, \omega_0) &= \int_{-\infty}^{\infty} \frac{dk_y}{[(k_x^2 + k_y^2) D\tau_\theta + 1]^2 + (\omega_0 \tau_\theta)^2} = \\ &= \frac{\pi}{\sqrt{2D\tau_\theta}} G_\theta(k_x \sqrt{D\tau_\theta}, \omega_0 \tau_\theta) \end{aligned}, \qquad (74)$$

$$G_\theta(x_1, a_1) = \frac{1}{[(x_1^2+1)^2 + a_1^2]^{1/2} \left\{[(x_1^2+1)^2 + a_1^2]^{1/2} + x_1^2 + 1\right\}^{1/2}}, \qquad (75)$$

$$x_1 = k_x \sqrt{D\tau_\theta}, \; a_1 = \omega_0 \tau_\theta \qquad (76)$$

One can see from (75) that in the case of internal wave breaking spectrum (71) the spectrum $E_\theta(k_x)$ again has two regions of spatial scales. The region of large scales or small wave numbers, where $k_x << 1/\sqrt{D\tau_\theta}$ and $k_x << \sqrt{\omega_0 / D}$, is characterized by a slow decrease of the spectral density with growth of the wave number. The spectral width of this region increases with increase the parameter a_1. The second region is a region of small spatial scales or large wave numbers, where $k_x >> 1/\sqrt{D\tau_\theta}$ and $k_x >> \sqrt{\omega_0 / D}$. Then, as it follows from (75), the spectrum $E_\theta(k_x)$ takes the form

$$E_\theta(k_x) = <\theta>^2 \frac{\sigma_w^2 \ell_w^2}{8\pi h^2 D^2} k_x^{-3}. \qquad (77)$$

Thus the power law "k_x^{-3}" is valid for the 1-D spectrums E_θ (k_x) in the range of wave numbers $k_x >> \max\{1/\sqrt{D\tau_\theta}, \sqrt{\omega_0 / D}\}$ and. In this region of spatial scales, the spectrums of the ocean surface non-uniformities of the θ-fields do not depend on the characteristic time scales τ_θ and ω_0^{-1}, they depend on the diffusivity coefficient D and fall off with increase D proportionally with D^{-2}. Krasnopevtsev and Fedorov [30, 1978] have shown that the power law "k_x^{-3}" well represents surface temperature fluctuations in the range of spatial scales from several hundred meters to 10 km. Figure 10 shows the 1D-spectrums E_θ (k_x) in the dimensionless form $E_\theta (k_x) / E_\theta(0)$ (equation (69), w' is delta-correlated, the power law "k_x^{-1}") and as G_θ (x_1) (equations (73) – (76), spatial white noise at the frequency ω_0, the power law "k_x^{-3}").

The variance, σ_θ^2 for the spectral function, $\Phi_w(\boldsymbol{k},\omega)$ in the form (71) takes the following form

$$\sigma_\theta^2 = <\theta>^2 \frac{\sigma_w^2 \ell_w^2}{4\pi h^2 D\omega_0}\left(\frac{\pi}{2} - \arctan\frac{1}{\omega_0 \tau_\theta}\right), \tag{78}$$

which has two asymptotic representations due to the value of the temporal dimensionless parameter $\Omega = \omega_0 \tau_\theta$. In the case when $\tau_\theta << \omega_0^{-1}$ i.e. the time of degradation τ_θ is shorter than the temporal scale ω_0^{-1} of temporal variability of the field w, the variance σ_θ^2 is

$$\sigma_\theta^2 = <\theta>^2 \frac{\sigma_w^2 \ell_w^2 \tau_\theta}{4\pi h^2 D}, \tag{79}$$

As it is seen from the expression (79), the variance σ_θ^2 is an increasing linear function of the characteristic time τ_θ, and does not depend on the characteristic time ω_0^{-1} of temporal variability of the field w.

In the opposite case, when $\tau_\theta >> \omega_0^{-1}$, i.e. the time scale ω_0^{-1} of temporal variability of the field w is shorter than the utilization time τ_θ, the variance σ_θ^2 is

$$\sigma_\theta^2 = <\theta>^2 \frac{\sigma_w^2 \ell_w^2}{8h^2 D\omega_0}. \quad (80)$$

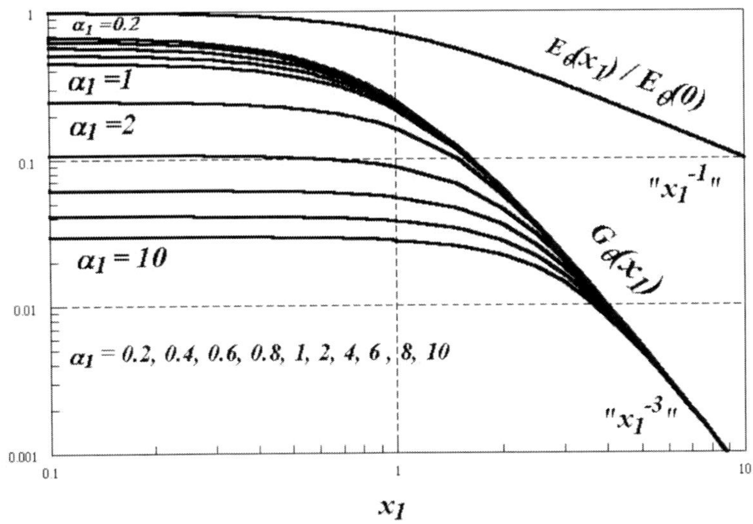

Figure 10. The 1D-spectrums $E_\theta(k_x)$ in the dimensionless form $E_\theta(k_x) / E_\theta(0)$ (equation (69), w' is delta-correlated) and as $G_\theta(x_1)$ (equations (73) – (76), spatial white noise at the frequency ω_0).

It is a linear function of the characteristic time ω_0^{-1}, and does not depend on the characteristic time τ_θ.

The discussed theory allows finding numerical estimates of the variance σ_θ^2 for realistic values of the quantities in equations (56) and (78). First we need to find values of the characteristic time τ_θ. For temperature, from equation (49), the quantity τ_T takes the order of magnitude of about ten days. In the case of biogenic salts, the time τ_C can be taken as of about the characteristic time of photosynthesis that takes the order of magnitude of about one day. The characteristic time ω_0^{-1} can be taken from measurements of turbulent bursts during TROPIC HEAT 1 at the Equator [Moum, et al., 1990], which give the estimate of the temporal variability of the field w as $\omega_0^{-1} \sim 10$ hours. Then the dimensionless parameter $\Omega_\theta = \omega_0 \tau_\theta$ takes the following values: for the upper layer temperature field it is $\Omega_T = \omega_0 \tau_T \approx 10^2$, and for the upper layer biogenic salts field it is $\Omega_C = \omega_0 \tau_C \approx 1-10$. Hence, the variances σ_θ^2 of both the temperature field and the biogenic salts field can be estimated using the equation (80).

Assuming that the turbulent energy of internal waves breaking regions has a uniform density of probability distribution, equal to $1/2\bar{k}_w$, where $\bar{k}_w$ is a mean value of k_w. Then the mean value $\overline{w}$ and the variance $\overline{w'^2}$ of the random field w are

$$\overline{w} = 0.8\kappa_4 \sqrt[4]{2k_e \bar{k}_w}\,, \tag{81}$$

$$\overline{w'^2} = \frac{\overline{w}^2}{24} = \frac{2}{75}\kappa_4^2 \sqrt{2k_e \bar{k}_w}\,. \tag{82}$$

The variance, σ_w^2, in Equation (66) can then be given in the following form

$$\sigma_w^2 = p(\frac{25}{24} - p)\overline{w}^2\,. \tag{83}$$

Now we need to express the quantity $p\overline{w}$ versus $<\theta>$ from Equation (56). Then the variance σ_θ^2 from the Equation (78) transforms to a new form, which is

$$\sigma_\theta^2 = \frac{1}{4\pi}\frac{(1-<\theta>)^2 \ell_w^2}{p\tau_\theta D}\left(\frac{25}{24} - 1\right)\frac{1}{\omega_0 \tau_\theta}\left(\frac{\pi}{2} - \arctan\frac{1}{\omega_0 \tau_\theta}\right). \tag{84}$$

This expression can be used to evaluate σ_θ^2 by the quantities obtained from measurements. The quantity $<\theta>$ can be taken as $<\theta> \sim 1$. In the case of the upper layer temperature field, the approximate equality $\overline{T}_s \approx T_a$ is permissible. For real ocean conditions we can assume $|\overline{T}_s - T_a| \sim 1^o C$, that is again $|T_a - T_\infty| \sim 10^o C$, and then $(1 - <\theta>) \sim 0.1$. The intermittency parameter p, according to [Grant et. al, 1968; Marmorino, 1990], takes the order of magnitude as $p \sim 10^{-2}$. The characteristic horizontal scale of turbulent regions ℓ_w can be taken as $\ell_w \sim 10^2\ m$ [Pozdenin, 1982]. The numerical value of the coefficient of horizontal turbulent diffusion D in the region of spatial scales from several hundred meters to 10 km is of the range $0.1 < D < 10\ m^2\ sec^{-1}$ [Okubo and Ozmidov, 1970]. Thus for the temperature field of the ocean upper layer, the standard deviation σ_T from (84) is $\sigma_T \sim 0.01 - 0.1°C$. These kind of the ocean upper layer temperature anomalies can be observed at the ocean surface at the appropriate spatial scales. The standard deviation σ_C of biogenic salt takes the order of magnitude $\sigma_C \sim 10^{-2}\ C_\infty$ for the same numerical values of the parameters.

CONCLUSION

As a result of atmospheric actions at the ocean surface, an upper turbulent layer in the ocean is formed close to the surface. It is separated from the ocean main interior by the pycnocline , the density jump layer, that is a physical lower boundary of the upper turbulent layer. The largest portion of the internal wave energy density is concentrated there. In the case of non-breaking internal waves, the turbulent upper layer reaches some dynamical and thermal equilibrium state with surface actions and practically does not interact with the main body of the ocean water below the pycnocline. This equilibrium changes when in the pycnocline, the seawater density jump layer, and other regions of stratified ocean water the internal waves break and create localized turbulent regions. There is a local mixing that is bounded in time by the lifetime of turbulence in those regions and spatially bounded by the sizes of these regions. The internal wave breaking mechanism constantly works in the ocean and forms the vertical transfer through the turbulent regions. The turbulent regions of the density jump layer interact with the upper layer turbulence, which drives the mean continuous vertical transfer between the upper layer and basic ocean interior.

As a result of this mechanism, there are the fluxes of heat and biogenic salts at the lower boundary of the upper layer, and the variability of horizontal non-uniformities of hydrophysical and hydrobiological fields at the ocean surface. The theoretical analysis of this phenomenon is carried out. Theoretical results are supported by available field measurements.

Thus the internal waves can collapse, forming turbulent regions in the pycnocline that cause the vertical transport of energy, heat, salinity and various substances between the upper layer and the ocean interior. The interaction of the upper layer turbulence and internal wave breaking turbulence can make an important contribution to the surface inhomogeneities of hydrophysical and biological fields. The numerical evaluations show that the present mechanism gives values of variability parameters which agree with the observations. Hence the internal wave breaking turbulence can play an important role in the ocean vertical structure formation and the transport of mineral nutrition to the upper layer.

REFERENCES

Barenblatt, G. I. (1977). "Intense Interaction of Gravity Waves and Turbulence", Ac. Sci. USSR, *Atmospheric and Oceanic Physics*, V. *13*, No. 8, pp. 845-849.

Barenblatt, G. I. (1978). "Dynamics of Turbulent Spots and Intrusions in a Stably Stratified Liquid", Ac. Sci. USSR, *Atmospheric and Oceanic Physics*, V. *14*, No. 2, 195-206.

Barenblatt, G. I. (1996). *Scaling, self-similarity, and intermediate asymptotics.* Cambridge University Press, 386.

Barenblatt, G. I. (1982). "*Spreading of a Turbulent Layer*", In: "N. Ye. Kochin and the Problems of Mechanics", Moscow, Nauka.

Barenblatt, G. I. & Benilov, A. Yu. (1982). "Influence of Internal Waves on Heterogeneity of the Hydrophysical Characteristics of Ocean Surface", In: "Interaction of Large-Scale Internal Waves on the Sea Surface", Gorky, *Inst. Appl. Phys. Ac. Sci. USSR*, 52-74.

Benilov, A. Yu. (1973). "Generation of Turbulence in the Ocean by Surface Waves", Ac. Sci. USSR, *Atmospheric and Oceanic Physics*, V. *9*, No. 3, 293-303.

Benilov, A. Yu. (1984). "*Formation of the Biogenic Fields in the Upper Layer of the Ocean by the Turbulent Mixing*", Moscow The Center of Co-ordination of COMECON (SEV), V. *18*, 69-77.

Benilov, A. Yu. (1986). "Hydrophysical Model of the Formation of the Biogenic Fields in the Upper Layer of the Ocean", In: "Investigation of Ecosystem of the Pelagic Zone of the Black Sea", Moscow, The Center of Co-ordination COMECON (SEV), 48-60.

Benilov, A. Yu. (1989). "On the Mixed Regions Collapse in the Ocean Pycnocline", *Ac. Sci. USSR, Oceanology*, V. *29*., No. 1, 33-39.

Benilov, A. Yu. (1989). "Low-Parameter Specification of Oceanic Ecosystem", In: "The Modeling Process in the Ocean", (Editors: Monin A. S., Seidov D. G., Vinogradov N. E.), Moscow, "Nauka".

Benilov, A. Yu. & Gavrilin, B. L. (1988). "Island Turbulence of the Ocean Pycnocline", *Ac. Sci. USSR, Oceanology*, V. *28*, No. 6, 700-705.

Benilov, A. Yu., Voropaev, S. I. & Zhmur, V. V. (1983). "Simulation of the Oceanic Upper Turbulent Layer During Heating", Ac. Sci. USSR, *Atmospheric and Oceanic Physics*, V. *19*, No. 2, 175-184.

Benilov, A. & Ly, L. N. (2002). Modeling of Surface Waves Breaking Effects in the Ocean Upper Layer. *Mathematical and Computer Modelling. 35*, 191-213.

Benilov, A. (2013). Air-Sea Interaction Ms: 68 Surface Waves. Encyclopedia of Atmospheric Science 2nd Edition, *Elsevier, 16.*

Bock, E. J., Carlson, D. J., Moum, J. N. & Cullen, J. J. (1990). "Characterization of Hydrodynamic and Chemical Slicks by Simultaneous Measurements of Time-Domain Surface Wave Spectra, Chemical and Biological Surface Enrichments, and Near-Surface Physics", *EOS*, V. *71*, No. 2, p. 72.

Browand, F. K. & Winant, C. D. (1973). "Laboratory Observations of Shear-Layer Instability in a Stratified Fluid", *Boundary Layer Meteorology*, V. *5*, No. 1/2, 67-77.

Curry, J. A. & Webster, P. J. (1999). *Thermodynamics of Atmospheres and Oceans*. Academic Press, 467.

Dickey, T. D. (1991). "The Emergence of Concurrent High-Resolution Physical and Bio-Optical Measurements", *Reviews of Geophysics*, V. *29*., No. 3, 383-413.

Fedorov, K. N. & Sklyarov, V. Ye. (1980). "Prospects for Studying the Ocean with the Help of Artificial Earth Satellite's", In: Khodarev, Yu. K., and Kravtsov, V. I. (Editors), "Investigation of Earth Resources by Space Vehicles", Part 2, Moscow, *Ac. Sci. USSR*, 42-57.

Fedorov, K. N. (1981). "Physical Structure of the Near-Surface Layer of the Ocean", *Meteorology and Hydrology*, No. *10*, 58-66.

Frankingnoul, C. & Hasselmann, K. (1977). "Stochastic Climate Models: Part 2. Application to Sea-Surface Temperature Anomalies and Thermocline Variability", *Tellus*, V. *29*, 289-305.

Fairbridge, R. W. (Editor), (1966). *The encyclopedia of oceanography*, (Encyclopedia of earth sciences series), Reinhold Pub. Co. 1021.

Frost, W. & Moulden, T. H. (1977). *Handbook of Turbulence*, New York, Plenum Press, 498.

Galaziy, G. I., Shifrin, K. S. & Sherstyankin, P. P. (1979). (Editors), "*Optical Methods of Studying Oceans and Inland Water Basins*", Novosibirsk, Nauka, 373.

Glazman, R. E. (1991). "Statistical Problems of Wind-Generated Gravity Waves Arising in Microwave Remote Sensing of Surface Winds", *IEEE Transactions on Geoscience and Remote Sensing*, V. *29*, No. 1, 135-142.

Grant, H. L., Moilliet, A. & Vogel, W. M. (1968). "Some Observations of the Occurrence of Turbulence in and above the Thermocline", *J. Fluid Mech.*, V. *34*, Part 3, 443-448.

Hasselmann, K. (1966). "Weak-Interaction Theory of Ocean Waves", In: Holt, M. (Editor), "*Basic Developments in Fluid Dynamics*", V. *2*, Academic Press, New-York, London, 5-57.

Hoffman, K. (1989). "Computational Fluid Dynamics for Engineers". Engineering Education System, Austin, Texas, 567.

Hood, R. R., Abbott, M. R., Huyer, Adr. & Kosro, P. M. (1990). "Surface Patterns in Temperature, Flow, Phytoplankton Biomass, and Species Composition in the Coastal Transition Zone of Northern California", *Journal of Geophysical Research*, V. *95*, No. C10, 18081-18094.

Ivanenkov, V. N. (1979). "The General Knowledge on Nitrogen, Phosphorus, and Silicon. In: "*The Ocean Chemistry*", V. *1*, "The Ocean Water Chemistry", Moscow, "Nauka", 176-186.

Ivanenkov, V. N. (1979). "The General Laws of a Distribution of the Biogenic Salts in the World Ocean", In: "*The Ocean Chemistry*", V. *1*, "The Ocean Water Chemistry", Moscow, "Nauka", 188-229.

Ivanov, A. V., Ostrovskiy, L. A., Soustova, I. A. & Tsimring, L. Sh. (1982). "Interaction of Internal Waves and Turbulence in the Upper Ocean Layer", In: "Interaction of Large-Scale Internal Waves on the Sea Surface", Gorky, *Inst. Appl. Phys. Ac. Sci.* USSR, 75-84.

Kanta, L. & Clayson, C. (2000). *Small Scale Processes in Geophysical Fluid Flows*. Academic Press, 883.

Kelly, K. & Caruso, M. (1990). "A Modified Objective Mapping Technique for Scatterometer Wind Data", *J. Geophys. Res.*, V. *95*, No. 8, 13483-13496.

Kitaygorodskiy, S. A. (1970). "*Physics of Interactions Between Atmosphere and Ocean*", Leningrad, Gidrometeoizdat, 284.

Krasnopevtsev, A. Yu. & Fedorov, K. N. (1978). "On the Thermo-Halo Non-Uniformities of Spatial Structure of the Ocean Upper Layer", In: "Analysis of Variability of Physical Processes at the Ocean", Moscow, *Inst. Oceanology Ac. Sci.* USSR, 59-73.

Kraus, E. B. & Businger, J. A. (1995). "*Atmosphere – Ocean Interaction*". Oxford University Press, 2nd edition, 362.

Kuznetsov, O. A. (1978). "Role of High Frequency Gravity Waves in Aerodynamics Resistance of the Sea Surface", Ac. Sci. USSR, *Oceanology*, V. *18*, No. 2, 203-207.

Lalbeharry, R., Khandekar, M. L. & Peteherych, S. (1990). "Wind Specification Based on Seasalt a Satellite Scatterometer and Conventional Winds for Driving on Ocean Wave Model", *J. Geophys. Res.*, V. *95*, No. C1, 761-773.

Longuet-Higgens, M. S. (1969). "On the Wave Breaking and the Equilibrium Spectrum of Wind-Generated Waves", *Proc. Roy. Soc.*, London, V. *A3 10*, No. 1501, 151-159.

Martin, S. (2011). *An Introduction to Ocean Remote Sensing*", Cambridge, 476.

Marmorino, G. (1990). "Turbulent Mixing in a Salt Finger Staircase", *J. Geophys. Res.*, V. *95*, No. C8, 12983-12994.

Minnett, P. J. (1990). "The Regional Optimization of Infrared Measurements of Sea Surface Temperature From Space", *J. Geophys. Res.*, V. *95*, No. C8, 13497-13510.

Monin, A. S. & Piterbarg, L. I. (1977). "Statistical Description of Internal Waves". *DAN USSR*, V. *234*, No. 3, 58-66.

Monin, A. S. & Ozmidov, R. V. (1985). "*Turbulence in the Ocean*". Springer, 264.

Monin, A. S. & Yaglom, A. M. (1987). Statistical Fluid Mechanics: Mechanics of Turbulence. Vol. *1*. (769), Vol. 2. (874), The MIT Press.

Moum, J., Herbert, D., Paulson, C. & Caldwell, D. (1990). "The Biggest Burst: Breaking Internal Waves at the Equator", *EOS*, V. *71*, No. 2, 102.

Naudascher, E. (1965). Flow in the Wake of Self-Propelled Bodies and Related Sources of Turbulence. *J. Fluid Mech.*, V. *22*, No. 1, 625-656.

Okubo, A. & Ozmidov, R. V. (1970). "Empirical Scale Dependence of Coefficient of Horizontal Turbulent Diffusion in the Ocean", Ac. Sci. USSR, *Atmospheric and Oceanic Physics*, V. *6*, No. 5, 534-536.

Ostrovskiy, A. G. & Piterbarg L. I. (1997). *Advection and Diffusion in Random Media*, Kluwer, 1997, 344.

Ostrovskiy, L. A. & Soustova, I. A. (1977). "The Upper Mixed Layer of the Ocean as a Sink of the Energy of Internal Waves", Ac. Sci. USSR, *Oceanology*, V. *17*, No. 6, 973-981.

Ostrovskiy, L. A., Soustova, I. A. & Tsimring, L. Sh. (1981). "Effect of Internal Waves on Small-Scale Turbulence in the Ocean", Gorky, *Inst. Appl. Phys. Ac. Sci. USSR*, Preprint No. 31, 15.

Pozdenin, V. D. (1982). "Some Characteristics of Generation and Structure of the Turbulent Equatorial Regions of the Pacific Ocean", Ac. Sci. USSR, *Oceanology*, V. *22*, No. 3, 363-368.

Schluessel, P., Emerty, W. J., Grassal H. & Mammen, Th. (1990). "On the Bulk-Skin Temperature Difference and Its Impact on Satellite Remote Sensing of Sea Surface Temperature", *J. Geophys. Res.*,V. *95*, No. C8, 13341-13356.

Schooley, A. N. & Stewart, R. W. (1963). "Experiments with a Self - Propelled Body Submerged in a Fluid with a Vertical Density Gradient", *J. Fluid Mech.*, V. *15*, 88-96.

Seomina, G. I. (1977). "*The Phytoplancton*", In: "Oceanology Ocean Biology, V. *1*. The Ocean Biological Structure", Moscow, "Nauka", 117-124.

Steele, J. H., Thorpe, S. A. & Turekian, K. K. (Editors), (2009). Encyclopedia of Ocean Sciences, 2nd Edition, 6-volume set, Academic Press.

Stewart, R. H. (1985). *Methods of Satellite Oceanography*. Univ. of Calif. Press, Berkley, 352.

Wahl, D. D. & Simpson, J. J. (1990). Physical Processing Affecting the Objective Determination of Near-Surface Velocity From Satellite Data", *J. Geophys. Res.*, V. *95*, No. C8, 13511-13528.

Wijesekera, H. W. & Willon, T. M. (1990). "Internal Waves and Mixing in the Upper Equatorial Pacific", *EOS*, V. *71*, No. 2, 102.

Willard, J. & Pierson, Jr. (1990). "Examples of Reasons for and Consequences of the Poor Quality of Wind Data From Ships for the Marine Boundary Layer: Implications for Remote Sensing", *J. Geophys. Res.*, V. *95*, No. C8, pp. 13313-13340.

Yefimov, V. V. & Khristoforov, G. N. (1971). "Wave and Turbulence Components of the Velocity Spectrum in the Upper Layer of the Ocean", Ac. Sci. USSR, *Atmospheric and Oceanic Physics*, V. *7*, No. 2, 200-211.

Zagorodnikov, A. A. (1978). "*Radar Survey of Sea Disturbance by Airborne Platforms*", Leningrad, Gidrometeoizdat, 293.

Zaslavskiy, M. M. (1981). "The Eigenvalue Relation's Characteristics of Wind Waves", Ac. Sci. USSR, *Atmospheric and Oceanic Physics*, V. *17*, No. 1, 59-66.

Zaslavskiy, M. M. & Leykin, I. A. (1981). "Interpretation of Measurements of the Eigenvalue Relation's Characteristics of Wind Waves", Ac. Sci. USSR, *Atmospheric and Oceanic Physics*, V. *17*, No. 6, 639-646.

In: Seawater
Editor: Mathew R. White

ISBN: 978-1-62808-606-5

Chapter 2

TRACE ELEMENTS IN SEAWATER IN *OFFSHORE OIL PRODUCTION PLATFORM* AREAS

***Sergio Felipe JerezVeguería*[1,2] *and José Marcus Godoy*[2,3]**

[1]Departamento de Química Analítica, Universidade Federal Fluminense, Outeiro de São João Batista s/n Campus do Valonguinho, Centro, CEP. Niterói, RJ, Brazil

[2]Departamento de Química, Pontifícia Universidade Católica do Rio de Janeiro, Rua Marquês de São Vicente, Gávea, CEP, Rio de Janeiro, RJ, Brazil

[3]Instituto de Radioproteção e Dosimetria, Comissão Nacional de Energia Nuclear, Av. Salvador Allende s/n, Jacarepaguá, CEP, Rio de Janeiro, RJ, Brazil

ABSTRACT

Produced water is the largest wastewater discharge in offshore oil, gas exploration and production processes. This water consists of formation water, the water naturally present in the reservoir, and floodwater injected into the formation. In general, in offshore operations, seawater is injected into the reservoir to maintain pressure in the formation for oil transportation. The produced water, that usually has a higher salt concentration than seawater, is separated from the oil and then discharged to the surface water. In this chapter, the characterization and evaluation of the trace elements and natural radionuclides in seawater in the *Bacia de Campos*, or Campos Basin, oil field region, in Brazil, is presented. In the present work, a representative region of oil-bearing activity in the *Bacia de Campos*, an oceanic region of the oil-field offshore platforms, represented by two fixed production platforms, was studied. The element and natural radionuclide content in dissolved and particulate produced water and seawater around the effluent discharge sites were studied. Six samples were collected at three different depths (surface (5 m), middle (45 m) and bottom (95 m)) in the oceanic region, with the sampling points allocated at 250 m, 500 m and 1000 m from the platforms, totaling 18 samples for each platform. The determination of V, Mn, Co, Ni, Cu, Zn, Cd and Pb was performed by Inductively Coupled Plasma Mass Spectrometry (ICP-MS) after solid phase extraction using online and offline montages. Inductively Coupled Plasma Optical Emission Spectrometry (ICP-OES) was used to determine Ba, Fe, Zn, Cd, Ni, Cr, Cu, Ag and Al in produced water and Ba in seawater. Uranium and molybdenum determinations were performed by ICP-MS with an ultrasonic nebulizer. Using one liter of produced water and twenty liters of seawater, after radiochemical separation, ^{210}Pb was determined by beta counting of ^{210}Bi, and ^{226}Ra and ^{228}Ra were determined by gross alpha and beta counting rate using a proportional 10-channel low-level proportional counter. The element concentrations in the effluents were, in general, compatible with the literature values and below the maximum acceptable values established by the Brazilian legislation. The measured activity concentrations ranged from 0.8 to 6.0 Bq L^{-1} for ^{226}Ra, from 0.7 to 8.2 Bq L^{-1} for ^{228}Ra and from 0.04 to 1.9 Bq L^{-1} for ^{210}Pb, similar to values reported for other produced water around the world. A strong correlation between Ba and radium isotopes in produced water was observed. Element concentrations in seawater were similar at the two study sites. The values obtained for ^{226}Ra, ^{228}Ra and ^{210}Pb in seawater, were below 5 mBq L^{-1}. The results showed a high degree of dilution regarding the produced water samples for barium, radium isotopes, copper, cadmium, nickel and ^{210}Pb. In spite of the elevated element and natural radionuclide concentrations in the produced water, concentrations in the seawater from the vicinity of the effluent discharges were at the local background level.

1. INTRODUCTION

The production of oil and gas is generally accompanied by the emission of water, named produced water, which comprises the following: formation water (water naturally present in the reservoir), injection water (usually sea water previously injected into the reservoir to maintain pressure while oil is removed), and water condensed with light hydrocarbons, which occurs in some gas production cases. This produced water has water as its main component, and, to a lesser degree, the organic and inorganic substances involved in oil production. The chemical composition of waste water is complex and includes dispersed oil, dissolved hydrocarbons, organic acids, metals, radionuclides and traces of chemical products that are added in the separation production lines. Although these constituents are present in small concentrations, the volumes discharged into the sea are high. The amount of water produced at offshore oil platforms is typically several times greater than the actual oil volume, typically 2000 to 40000 m^3 day^{-1}. On the other hand, at offshore platforms, the volume of produced water is much lower; typically from 2 to 30 m^3 day^{-1} [1]. At the start of production, the amount of water produced along with oil is usually low, but the as the oil field ages, this volume may increase considerably [2].

1.1. Emission of Trace Elements in Offshore Oil Production

Formation water, which has been in contact with geological formations for millions of years, contains a wide variety of dissolved inorganic salts and organic substances characteristic of the geological formation where the water is produced. The total concentration of dissolved salts in produced water ranges from 3g L^{-1} until close to saturation, but is usually higher than in seawater. Although it resembles concentrated seawater, produced water has a slightly different ionic composition, even though it presents sodium and chloride as the most abundant ions. Like sea water, produced water contains trace metals and organic substances, as well as dispersed oil [3].

The major inorganic components in produced water are inorganic salts, trace metal and nutrients. They are, as a rule, similar to those in seawater. However, formation water usually has a higher salt concentration and a lower pH than seawater.

The use of seawater in the extraction process causes a progressive dilution of the formation water which will, in time, cause the produced water to become increasingly closer to seawater in its composition and characteristics. Several metals, such as barium, nickel, lead, chromium, silver and zinc are present in produced water at substantially high concentrations [3]. Different studies regarding the chemical characterization of produced water from wells in the North Sea, Gulf of Mexico and the coast of Canada have shown considerable variability that can be attributed, in part, to the different geological characteristics of the reservoirs.

Other factors may also be relevant, i.e. Neff et. al. [4] have suggested that the corrosion of galvanized piping and equipment could be the source of zinc and lead in the produced water.

On the other hand, nickel and vanadium are porphyrin constituents present in significant concentrations.

1.2. Emission of Natural Radionuclides in Offshore Oil Production

Natural radionuclides in the production processes of oil and gas are described as Naturally Occurring Radioactive Material (NORM), with ^{226}Ra (half-life of 1602 years, natural ^{228}U series), ^{228}Ra (half-life of 5.8 years, natural ^{232}Th series), ^{210}Pb (half-life of 22.3 years, natural ^{228}U series) and ^{222}Rn (half-life of 3.8 days, natural ^{228}U series) being the most important [5]. High activities of ^{226}Ra and ^{228}Ra are often discharged by the oil and gas industry, particularly in produced water.

Uranium and thorium are part of the rock matrix, and are essentially insoluble in reservoir fluids. On the other hand, radium is a water soluble element and is therefore leached and mobile. Like barium and strontium, radium belongs to the Group IIA of the periodic table and has similar chemical properties.

In the extraction process, the contact of formation water with the seawater that is injected into the wells leads to a disruption of the chemical equilibrium, forming $BaSO_4$ and $SrSO_4$ precipitates, which co-precipitate radium isotopes. A portion of radium is deposited during scale formation within the oil transport pipe, while the remaining portion is discharged into the sea with the effluent resulting from the oil-water separation units [5-7].

In this chapter, the results of an environmental study conducted at the *Bacia de Campos* area, Brazil are presented. The concentrations of trace metals Pb, Cu, Ni, V, Zn, Cd, Ba, As, Se, Mn, Co, U and ^{226}Ra, ^{228}Ra and ^{210}Pb in produced water from oil production platforms and in seawater (filtrate and particulate material) in the discharge areas, as well as the dispersion of trace elements and natural radionuclides in seawater, were assessed. The environmental impact due to the activity of oil exploration in the region was also evaluated.

2. Study Area

The activity of the offshore oil industry at the *Bacia de Campos* area occupies an extensive area of 40,000 km^2, ranging from Cabo Frio to Cabo de São Tomé (latitudes 21°30'S and 23°00'S) and from the shoreline (longitude 39°30'W) to the oceanic region (longitude 42°30'W) of the oil fields, in water depths ranging from 80 to 2000 meters. This entire region is susceptible to the contribution of various pollutants from both the economic activity of the coast and the oil exploration and production at the oil platforms. Due to the large size of the *Bacia de Campos* area, two oil production Platforms (*Pampo* and *Pargo*) were studied, located between the isobaths of 90 and 300 meters and latitudes of 22° 23° 30'S, where both the discharge contributions from the platforms to the marine environment and the prospect of these discharges reaching the coastal region were assessed.

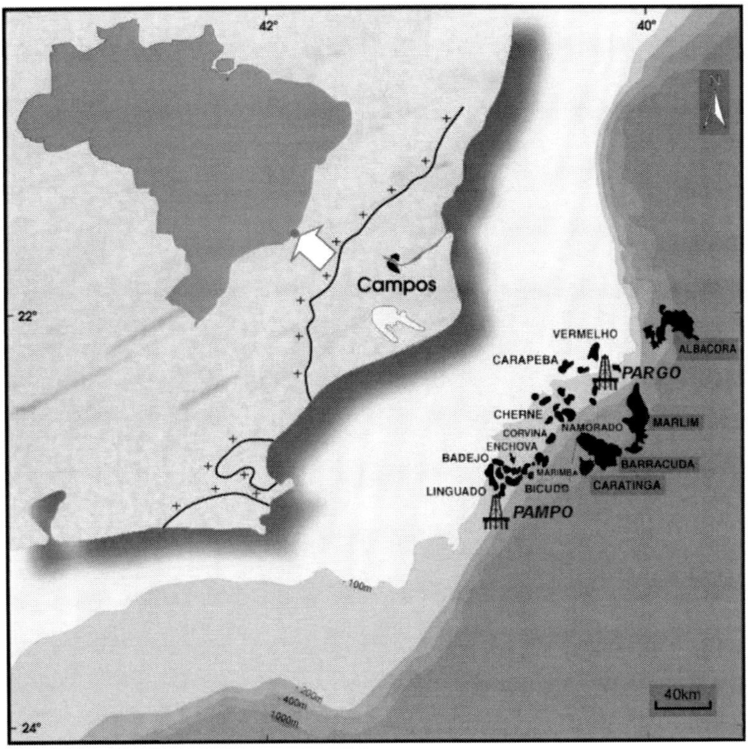

Figure 1. Map of the *Bacia de Campos* offshore oil field showing the location of the *Pargo* and *Pampo* platforms.

3. Experimental Procedures

3.1. Sampling Sites and Field Methods

The produced water was collected in 1 L glass vials, previously treated with HNO_3 1.0 mol L^{-1}, at the effluent discharge points from the oil-water separation units of the studied platforms. The samples were filtered with Whatman 42 filters. The filtrates were used for the analysis of dissolved elements and the filters for the analysis of particulate matter.

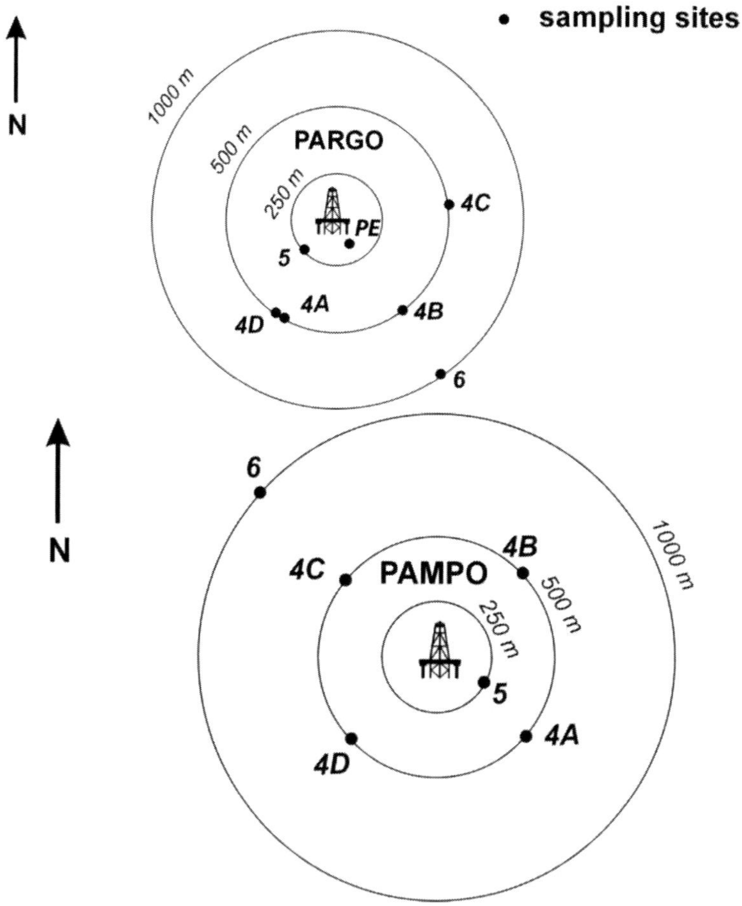

Figure 2. Location of the seawater sampling points around the *Pargo* and *Pampo* platforms.

Concerning the collection of seawater surrounding the platforms, a sample design in a cross shape was established, as suggested by Carney [8], where the sampling stations were allocated, with regard to the platform, at 250, 500 and 1000 meters, in a northeast-southwest alignment, which is the preferential direction of the Brazil current at the site, and 250 and 500 m in the northwest-southeast direction. The sampling points were established based on the preferential direction of the current at the sampling moment, and sampling was conducted at three depth levels: surface (≈ 5 m), middle (≈ 45 M) and near the bottom (≈ 95 m). Figure 2 shows the location of the sampling points at the *Pampo* (22°45'48''S, 40°45'14''W) and *Pargo* (22°15'16''S, 40°19'51''W) platforms at the time of seawater sampling [9]. Another sampling point, at the effluent discharge site at the *Pargo* platform, was included (PE).

Samples were obtained using a Go Flo bottle (General Oceanics) internally coated with Teflon. The seawater samples were transferred to 100 mL high density polyethylene bottles and kept at -10° C until the analyses. 20 L of seawater were collected with a Niskin bottle (General Oceanics) and filtered through 0.45 μm cellulose acetate Millipore filters for the determination of radionuclides and metals associated to the suspended material. At the *Pargo* platform, radium was additionally assessed in six sampling stations by preconcentration from 400 L superficial seawater (≈ 5 m), percolated at a rate of 10 L min^{-1} through two cellulose cartridges (25 cm x 6.2 cm), coated with MnO_2 and placed in series into a standard water filtering cartridge holder [10]. After pre-concentration, the MnO_2-cellulose cartridges were placed in plastic bags, sealed and identified.

3.2. Analytical Methods for the Determination of Trace Elements

3.2.1. Determination of Ba, Fe, Zn, Cd, Ni, Cr, Cu, Ag and Al in Soluble Fraction by ICP-OES

Concentrations of Ba, Fe, Zn, Cd, Ni, Cr, Cu, Ag and Al in produced water and Ba in seawater were obtained by direct sample measurement in an inductively coupled plasma optical emission spectrometer (ICP-OES) Perkin Elmer, model Plasma 1000, under the conditions shown in Table 1. For Ba in seawater, direct measurements were made and a calibration curve was constructed from a barium standard diluted in HNO_3 1% v/v, using 2.5, 5.0 and 10.0 mg L^{-1} solutions. In order to evaluate the matrix effect, two standard curves were constructed: one in HNO_3 1% v/v and a standard addition curve.

The ratio between the linear coefficients of the two curves was 1.2. This value was used to correct the measured concentrations of the other seawater samples. Milli-Q water was used as the blank. To improve the efficiency and repeatability of the analyses, Sc was introduced as an internal standard, with the aim of correcting interferences due to the electronic instability of the device, changes in the aspiration rate of the sample, nebulization and matrix effects due to the high salt content in seawater.

Table 1. Analytical and operating conditions of the ICP-OES

Rf power	1000 W
Nebulizer Ar flow	1.0 L min^{-1}
Plasma Ar flow	15.0 L min^{-1}
Auxiliary Ar flow	1.0 L min^{-1}
Nebulizer	Cross-flow
Nebulization chamber	Scott-Fassel
Sample aspiration rate	1.0 mL min^{-1}
Photomultiplier voltage	675 V
Calibration	External with Sc as internal standard
Replicates	3
Wavelength (nm)	Ba 455.403; Fe 273.955; Zn 213.856; Cd 228.802; Ni 231.604; Cr 267.716; Cu 324.754; Ag 238.289; Al 308.215

Table 2. Parameters of the analytical curves obtained by ICP-OES for the analysis of produced water

Element	Concentration range (mg L^{-1})	Sensitivity Intensity (mg $L^{-1})^{-1}$	R	3σ-LOD (mg L^{-1})
Al	0.01 – 0.05	1398	1.0000	0.027
Cr	0.01 – 0.10	3063	0.9959	0.004
Fe	0.01 – 1.00	2594	0.9994	0.002
Ni	0.05 – 0.10	979	0.9985	0.024
Cu	0.01 – 0.10	4503	0.9972	0.003
Zn	0.01 – 1.00	6112	0.9985	0.002
Ag	0.01 – 0.05	3516	0.9957	0.018
Cd	0.01 – 0.05	2659	0.9963	0.003
Ba	1.00 – 10.00	2858	0.9997	0.0002

Figure 3 exemplifies the standard curves obtained in this study for the determination of the elements of interest in produced water. The curves showed good linearity, resulting in a correlation coefficient of R> 0.996 for most elements. Table 2 shows the parameters of the curves. Cd, Ni, Cr, Al and Ag concentrations were below the limit of detection (LOD).

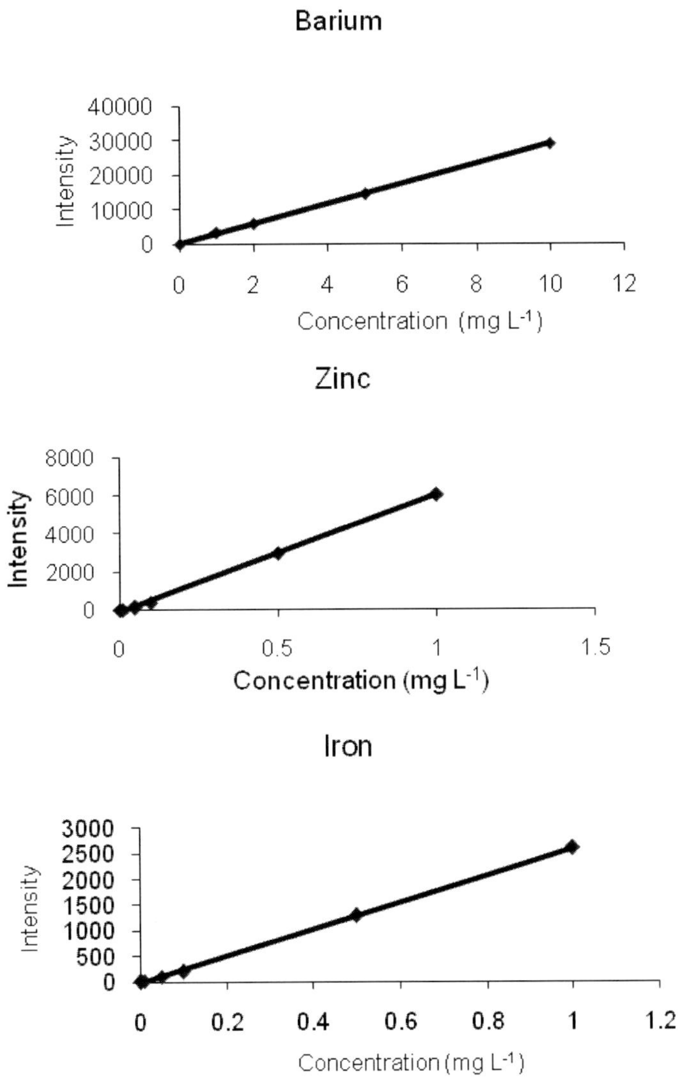

Figure 3. (Continued)

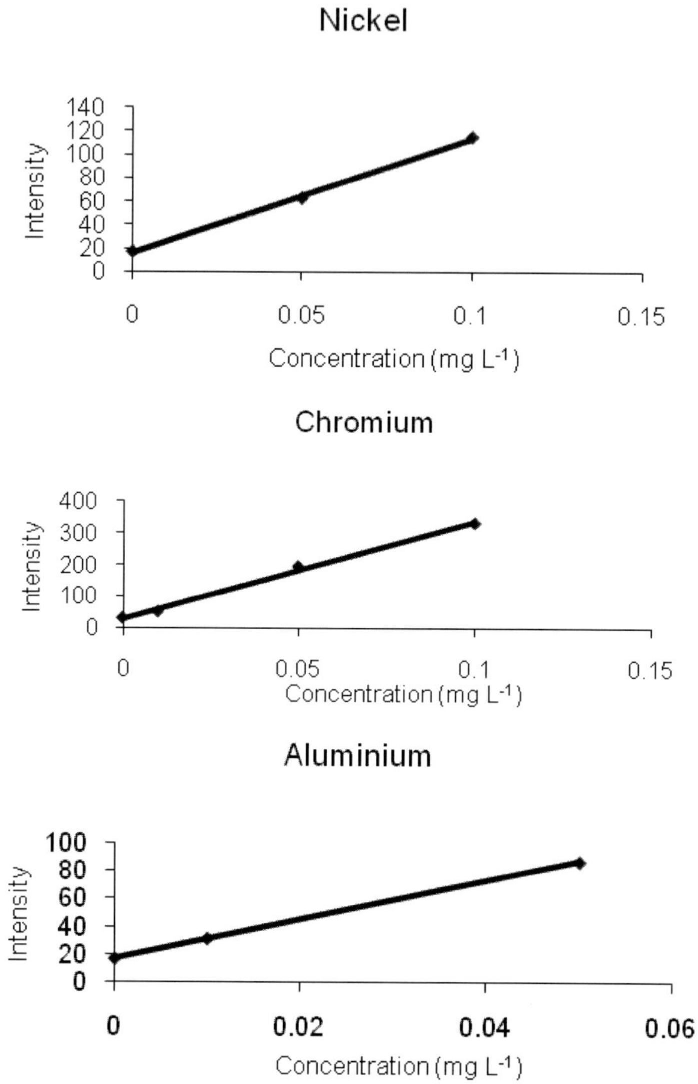

Figure 3. Analytical curves obtained by ICP-OES for the determination of Ba, Zn, Fe, Ni, Cr and Al in produced water.

The correlation coefficient of the curve for Ba in seawater was 0.9996 and the limit of detection was 0.25 μg L^{-1}, showing the suitability of this method for this study. Although the value of barium in the certified CASS-3 sample is not certified, the observed value in the present study was of 6.0 ± 0.6 μg L^{-1}.

3.2.2. Direct Determination of U and Mo in Soluble Fraction by ICP-MS with an Ultrasonic Nebulizer

The technique of ICP-MS with an ultrasonic nebulizer (USN-ICP-MS) was employed for the determination of U and Mo. A Perkin-Elmer Sciex ELAN-5000 ICP-MS, operated in quantitative mode (external calibration) using ^{205}Tl as internal standard was used.

Table 3 presents the analytical and operation conditions of the ICP-MS. The conventional nebulizer (cross-flow) was replaced by an ultrasonic nebulizer (CETAC U-6000 AT), for increased sensitivity. For the same reasons, a Pt sampler/skimmer pair instead of Ni was used. The analytical solutions in HNO_3 1% v/v were used to construct calibration curves (0.05, 0.25, 2.5, 5.0 and 10.0 μg L^{-1}), prepared by the dilution of a 1000 mg L^{-1} standard stock solution. For the analysis, the seawater samples were diluted 1:100 with Milli-Q water, acidified with sub-distilled HNO_3 and ^{205}Tl was added as the internal standard. The reference sample, coastal water CASS-3, was used to validate the developed methodology. V was also determined following this same method, after sample dilution of 1:100 with Milli-Q water, and the addition of ^{45}Sc as standard, under the conditions shown in Table 3. A calibration curve from a standard vanadium solution in HNO_3 1 % v/v at 0.05, 0.25, 2.5, 5.0 and 10.0 μg L^{-1} was constructed. HNO_3 1% v/v containing scandium at the same concentration as the curve and the samples was used as blank. The ^{51}V isotope (99.76% isotopic abundance) was used for the determinations. Due to the possible polyatomic interference of $^{35}Cl^{16}O$ and due to the high salinity of the samples even after a 100-fold dilution, it was also used a column pre-concentration method with elimination of the saline matrix, as described below.

Table 3. Analytical and operating conditions of the USN-ICP-MS

Rf power	1000 - 1050 W
Nebulizer Ar flow	1.0 L min^{-1}
Plasma Ar flow	15.0 L min^{-1}
Auxiliary Ar flow	1.0 L min^{-1}
Nebulizer	Ultra-sonic
Sample aspiration rate	1.0 mL min^{-1}
Scan mode	*peak hopping* and normal resolution
Dwell time	100 ms
Calibration	External, with internal standards: ^{45}Sc, ^{205}Tl.
Replicates	3
Isotopes	^{51}V, ^{98}Mo, ^{238}U

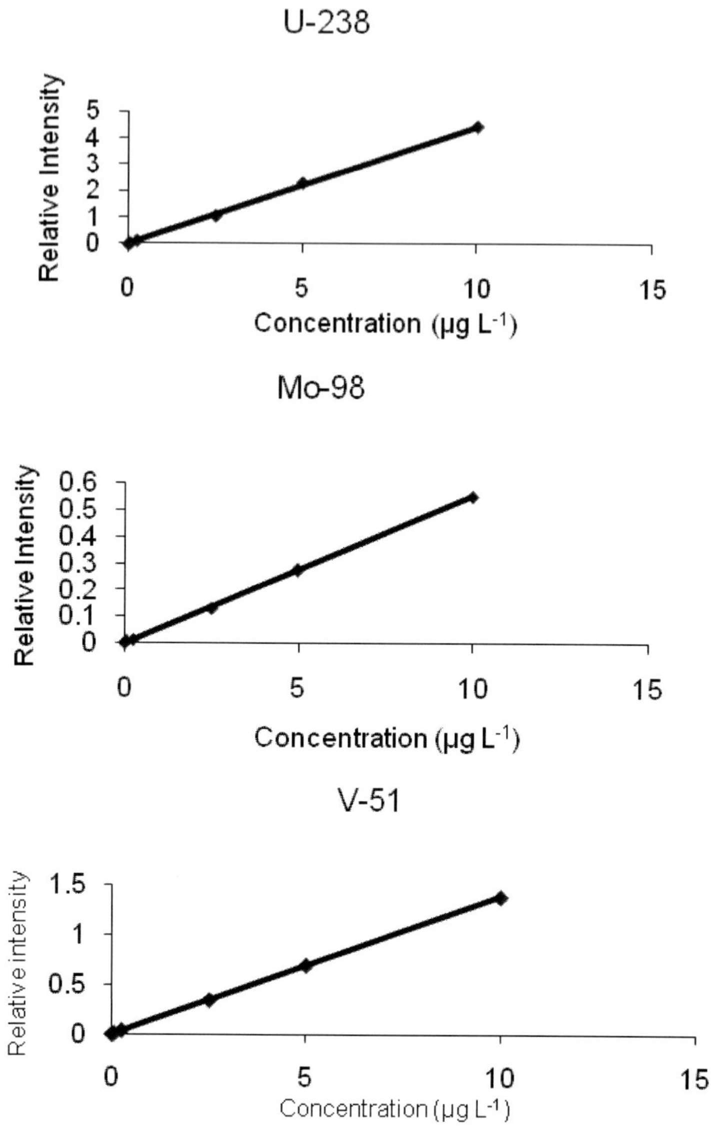

Figure 4. Calibration curve for U, Mo and V by ICP-MS with an ultrasonic nebulizer.

Given the natural isotopic composition of uranium, its mass/concentration corresponds almost exclusively to isotope ^{238}U (99.27%) and, due its high half-life (4.51×10^9 years), its specific activity makes methods based on mass determination competitive with those based on radioactivity measurements.

The ICP-MS with an ultrasonic nebulizer technique was adequate for the determination of U, V and Mo in produced water and seawater. Figure 4 shows the analytical curves.

The curve parameters are shown in Table 4. The limits of detection of the method allowed for sample dilution at a 1:100 ratio, minimizing signal suppression effects. The reference sample CASS-3 was used to validate the proposed methodology, with good results. Although the uranium content of sample CASS-3 is considered "informative only", the mean value was 2.7 ± 0.6 μg L^{-1}, consistent with the reported value.

The results demonstrate that this technique seems to be ideal for the determination of uranium and molybdenum in seawater, not only because of the excellent limit of detection (0.06 and 1 μg L^{-1}) but also because of the absence of spectral interferences in this mass range. In turn, V values were consistent with those reported in the literature (Table 5).

Table 4. ICP-MS with an ultrasonic nebulizer analytical curve parameters for the analysis of seawater

Element	Concentration range (μg L^{-1})	Sensitivity (cps L $μg^{-1}$)	R	3σ-LOD* (μg L^{-1})
^{51}V	0.5 – 10	0.1383	1.0000	0.002
^{98}Mo	0.5 – 10	0.055	0.9999	0.011
^{238}U	0.5 – 10	0.4464	0.9994	0.0006

*Multiply by the dilution factor to obtain the sample LOD.

Table 5. Results of the CASS-3 certified seawater analysis by ICP-MS with ultrasonic nebulizer

Element	Measured (μg L^{-1}) (n = 5)	Certified (μg L^{-1})
V*	1.40 ± 0.39	*n.c.*
Mo	9.20 ± 0.74	8.95 ± 0.27
U	2.70 ± 0.59	2.84**

* Measured V values in CASS-3: 1.18 ± 0.05 μg L^{-1} [11]; 1.50 ± 0.15 μg L^{-1} [12]; 1.47 ± 0.22 μg L^{-1} [13].

** Informed value.

n.c. not certified.

3.2.3. Determination of As and Se in Soluble Fraction by ICP-MS with Hydride Generation

The seawater matrix, abundant in elements such as Na, Ca, Cl and S, causes the formation of polyatomic species such as $^{40}Ar^{35}Cl$ which interferes with the determination of As, a monoisotopic element, with m = 75 [14-18]. To avoid this interference, it was opted for the determination of As by hydride generation. Thus, As and Se were determined by hydride generation coupled to inductively coupled plasma mass spectrometry (HG-ICP-MS) under the conditions shown in Table 6. The samples were pretreated with a pre-reduction process.

To convert all the inorganic arsenic to As (III), 1 mL of a stock solution of the pre-reducer agent (5% w/v KI, and ascorbic acid) and 3 mL of concentrated HCl were added to a 10 mL sample. After 30 minutes, the mixture was diluted with Milli-Q water to 25 mL. To convert all the inorganic selenium to Se (IV), 10 mL of concentrated HCl was slowly added to 10 mL of the sample, and the mixture was allowed to boil for 10 minutes. The samples were also diluted to 25 mL with Milli-Q water. ^{75}As (100% isotopic abundance) and ^{82}Se (19.9% of isotopic abundance) isotopes were used for the measurements. The analytical curves, with concentrations of 0.1, 0.5, 1 and 2 μg L^{-1} were constructed from 1000 mg L^{-1} stock solutions. Hydride generation occurred in a Perkin-Elmer FIAS-200 flow injection system, operating under the conditions prescribed by the manufacturer and using HNO_3 acid (3% v/v) as a carrier and $NaBH_4$ as the reducing agent at 0.5% w/v for As and 0.05% w/v for Se.

Table 6. Analytical and operational conditions of the HG-ICP-MS

Rf power	1200 – 1208 W
Nebulizer Ar flow	1.2 L min^{-1}
Plasma Ar flow	15.0 L min^{-1}
Auxiliary Ar flow	1.0 L min^{-1}
Nebulizer	*Meinhard*
Nebulization chamber	Cyclonic
Sample aspiration rate	1.0 mL min^{-1}
Scan mode	*Peak hopping* and normal resolution
Dwell time	100 ms
Calibration	External
Replicate number	3
Isotopes	^{75}As, ^{82}Se

Figure 5 presents the analytical curves. For arsenic, the LOD-3σ was 0.01 μg L^{-1} and for CASS-3, 1.15 ± 0.05 μg L^{-1}, with the certified value of 1.09 ± 0:06 μg L¯1, showing a good agreement between both values. For selenium, the LOD-3σ was 0.02 μg L^{-1} but this element in CASS-3 could not be quantified, since it is present in the same range as the method LOD (certified CASS-3 value: 0.020 ± 0.005 μg L^{-1}).

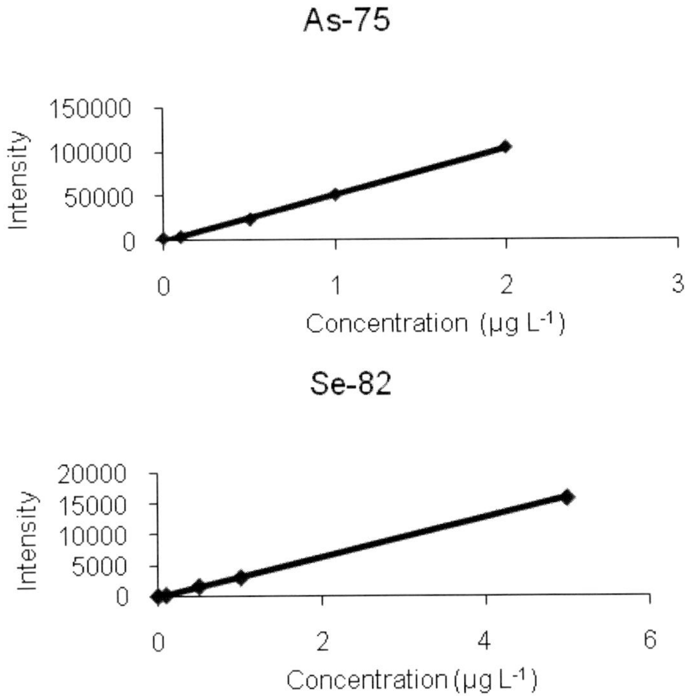

Figure 5. As and Se analytical curves by ICP-MS with hydride generation.

3.2.4. V, Mn, Co, Ni, Cu, Zn, Cd and Pb Determination in Soluble Fraction by ICP-MS with Matrix Separation

A metal pre-concentration method with simultaneous removal of the saline matrix was implemented [19]. The samples were analyzed using an ELAN 5000 ICP-MS (Perkin-Elmer-Sciex) operating in a quantitative mode and equipped with a concentric nebulizer and cyclonic spray chamber. The separation of the saline matrix and analyte pre-concentration was conducted in a micro-column packed with commercial iminodiacetate resin (Toyopearl ® AF-Chelate 650M). This procedure was implemented online using the FIAS

200 flow injection system coupled to ICP-MS (FI-ICP-MS). A summary of the operation parameters of the FI-ICP-MS is shown in Table 7.

Table 7. Operating conditions and data acquisition parameters for the FI-ICP-MS

Rf power	1200 W
Nebulizer Ar flow	0.95 L min^{-1}
Plasma Ar flow	15.0 L min^{-1}
Auxiliary Ar flow	1.0 L min^{-1}
Nebulizer	Concentric
Nebulizer chamber	Cyclonic
Aspiration rate	1.0 mL min^{-1}
Acquisition mode	Peak hopping
Dwell time	20 ms
Calibration	Analyte addition/external
Replicates	5
Readings per replicate	100
Isotopes	^{51}V, ^{55}Mn, ^{58}Ni, ^{59}Co, ^{60}Ni, ^{62}Ni, ^{63}Cu, ^{64}Zn, ^{65}Cu, ^{66}Zn, ^{114}Cd e ^{208}Pb

Table 8. Figures of merit of the FI-ICP-MS system obtained from analyte addition calibration curves in seawater at pH 8.2

Isotope	Sensitivity (cps L µg $^{-1}$)	R*	LOD† (µg L^{-1})
^{51}V‡	4881	0.9900	0.007
^{55}Mn	4527	0.9991	0.023
^{59}Co	3934	0.9966	0.005
^{58}Ni	2143	0.9954	0.056
^{60}Ni	884	0.9969	0.147
^{62}Ni	138	0.9985	0.102
^{63}Cu	1047	0.9919	0.057
^{65}Cu	511	0.9932	0.071
^{64}Zn	733	0.9902	0.057
^{66}Zn	363	0.9858	0.072
^{114}Cd	1069	0.9958	0,004
^{208}Pb	2470	0.9921	0.032

*Linear correlation coefficient; †calculated as the concentration equivalent to 3σ-standard deviation of the blank (n=10); ‡at pH 5.5.

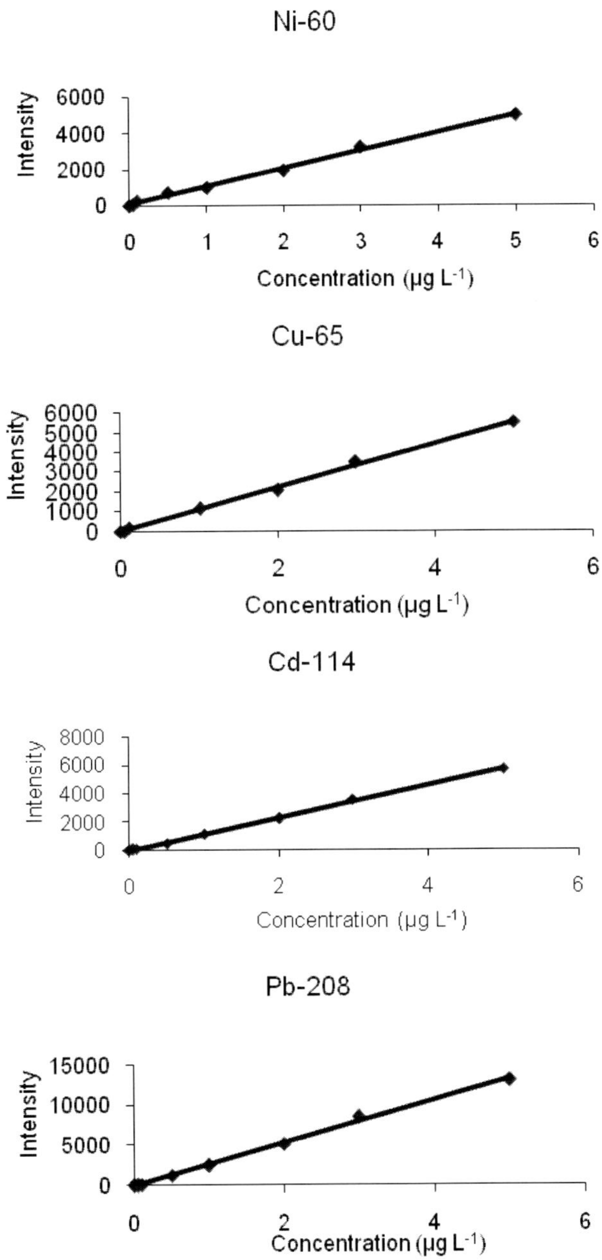

Figure 6. Analytical curves obtained by FI-ICP-MS for metal determination in seawater by online pre-concentration.

Micro-quantities of the standard solutions were used to prepare the analyte addition standard curves, to avoid changes in the seawater pH. The analytical curves were constructed ranging from 0.05 to 5 μg L^{-1}.

The methodology validation was conducted with a certified seawater sample, CASS-3. Due to its acidity (pH = 1.6) and the consequent difficulty in adjusting the pH to 8.2 in series, the validation was conducted using an ammonium-acetate buffer at pH 5.0. Table 9 shows the CASS-3 results for the elements of interest.

Table 9. Analytical results for CASS-3 using FI-ICP-MS at pH = 5.0

Element	Measured[a] (μg L^{-1})	Certified (μg L^{-1})
V	0.70 ± 0.07	*n.c.*
Mn[†]	2.51 ± 0.25	2.51 ± 0.36
Co	0.045 ± 0.001	0.041 ± 0.009
Ni	0.433 ± 0.061	0.386 ± 0.062
Cu	0.507 ± 0.177	0.517 ± 0.062
Zn	1.28 ± 0.19	1.24 ± 0.25
Cd	0.031 ± 0.009	0.030 ± 0.005
Pb	< 0.032	0.012 ± 0.004

[a]n = 9; *n.c.*: not certified; [†]pH=8.2.

Results showed good agreement between the certified and measured values for CASS-3 except for Mn, as expected, due to the poor recovery of this element at pH 5.0. In an independent CASS-3 analysis, conducted at pH 8.2, the Mn concentration found was 2.51 μg L^{-1}, demonstrating excellent agreement with the certified value.

3.2.5. Elements Determination in Particulate Material by ICP-MS

The cellulose acetate filters used for the seawater filtration and the paper filters used for the produced water filtration were transferred to Teflon™ vials with 3 mL of sub-distilled HNO_3, 2 mL of HF 40% v/v and 1 mL of 30% H_2O_2 v/v.

The vials were taken to a Microwave Accelerated Reaction System MARS 5 and the digestion was conducted with the following heating program: 5 min, 300 W (100%), 10 min, 600 W (66%); 5 min, 300 W (100%); 14 min, 600 W (92%); 2 min, 300% (0%). After cooling, the samples were dried on a heating plate at 90° C and the residue was dissolved in 2 mL of sub-distilled HNO_3

and brought up to a volume of 25 mL with Milli-Q water. 5 mL of these solutions were then analyzed by direct reading using the ICP-MS in the TotalQuant operating mode and with the use of ^{115}In and ^{205}Tl as internal standards. The operating conditions and mass/charge ratios employed are shown in Table 10.

Table 10. Operating and analytical ICP-MS conditions for metal determination in particulate material

Rf power	1000 W
Nebulizer Ar flow	1.0 L min^{-1}
Plasma Ar flow	15.0 L min^{-1}
Auxiliary Ar flow	1.0 L min^{-1}
Nebulizer	*Meinhard*
Nebulizer chamber	Cyclonic
Sample aspiration rate	1.0 mL min^{-1}
Scan mode	*Peak hopping* and normal resolution
Dwell time	100 ms
Calibration	TotalQuant with internal standards: ^{115}In e ^{205}Tl.
Replicates	3
Isotopes	^{27}Al, ^{51}V, ^{53}Cr, ^{57}Fe, ^{60}Ni, ^{65}Cu, ^{66}Zn, ^{75}As, ^{82}Se, ^{109}Ag, ^{114}Cd, ^{118}Sn, ^{138}Ba, ^{208}Pb e ^{238}U.

3.3. Analytical Methods for the Determination of ^{226}Ra, ^{228}Ra and ^{210}Pb

3.3.1. Radiochemical Analysis

Aliquots of 50 or 100 mL of produced water, depending on the Ba concentration previously determined, were then diluted with deionized water to one liter and used for the determination of dissolved radionuclides. For seawater, 20 L of filtered samples were used and for the particulate material, 20 mL of solutions resulting from the microwave digestion described above, were diluted to one liter too, and used in the analysis.

The radium isotopes (^{226}Ra and ^{228}Ra) and ^{210}Pb were determined following the methodology described by Godoy *et al.* [20]: 20 mg of Ba^{2+} and Pb^{2+} were added to the sample as carriers. Barium and lead sulfates were then precipitated by adding 50 mL of H_2SO_4 3 mol L^{-1}. The sample was then left to precipitate until the following day, when the supernatant was discarded and the precipitate washed with 20 mL of H_2SO_4 0.1 mol L^{-1} and centrifuged. Once the supernatant was discarded, the precipitate was dissolved in 2.0 g of nitrile triacetic acid (NTA), 40 mL of deionized water in an alkaline medium (7 mL of NaOH 6 mol L^{-1}) in a steam bath at 60 °C. Barium sulfate was re-precipitated with 5 mL of $(NH_4)_2SO_4$ (25 mg mL^{-1}), adjusting the pH to 4 - 5 with acetic acid. The precipitate $Ba(Ra)SO_4$ was separated for the radioisotope determination. In the filtrate, lead was precipitated as lead sulfide with 1 mL of Na_2S 1 mol L^{-1}. The supernatant was discarded and the sulfide dissolved by heating in 10 mL of HNO_3 5 mol L^{-1} at pH 4.5 - 5.0 adjusted with 40% ammonium acetate w/v. Finally, lead was precipitated as $PbCrO_4$ with 2 mL of Na_2CrO_4 30% w/v. The day and time of precipitation was noted and the precipitate was filtered with Whatman 42 paper filters, previously washed and weighed, already taking into account the geometry count of 2.5 cm in diameter. The precipitate was dried in a Petri dish at room temperature and weighed to gravimetrically determine the chemical yield. After a growth period of 30 days, the activity of ^{210}Pb was determined by the beta counting of its decay product, ^{210}Bi, in an EG&G Prof. Berthold LB770 low-background 10-channel low-level proportional counter. The $Ba(Ra)SO_4$ precipitate was mixed with 2.0 g of Na_2-EDTA and dissolved with 5 mL of concentrated NH_4OH and 25 mL of deionized water in a steam bath, followed by re-precipitation by the addition of concentrated acetic acid until pH 4.5 - 5.0. The day and time of precipitation was noted and, as for the lead precipitate, this precipitate was filtered, washed, dried and weighed. The activities of the radium isotopes were determined after 30 days (the necessary time to allow ^{226}Ra and ^{228}Ra daughters to reach radioactive equilibrium) by gross alpha counts for ^{226}Ra and gross beta counts for ^{228}Ra after covering the system with a paper filter (diameter = 2.8 cm) to shield the alpha particles, in the EG and G Prof. Berthold LB770 low-background 10-channel low-level proportional counter. The filters containing the precipitates were placed in stainless steel plates, 2.5 cm in diameter, and the count time was of 100 min. and 400 min. for alpha and beta, respectively. The typical chemical yield was 85 % and the measurement system was calibrated with standard sources of ^{226}Ra, ^{228}Ra and ^{210}Pb (4 Bq) provided by the Division of Radionuclide Metrology / National Laboratory of Metrology of Ionizing Radiation of the Institute of Radiation Protection and Dosimetry (IRD) of the National Commission of Nuclear Energy (CNEN).

3.3.2. Gama Spectrometry

This analytical technique [21] was used to directly determine ^{226}Ra and ^{228}Ra in the seawater samples collected around the *Pargo* platform by pre-concentration in MnO_2-cellulose cartridges, without the need for any radiochemical treatment. The cartridges were first dried at 80° C and subsequently incinerated in an oven at 500° C for 12 hours. The ashes were transferred to polyethylene bottles with the same geometry as the standards, resulting in an approximate mass of 60 grams. The vials were then capped and sealed and was waited a minimum of three weeks for the ^{226}Ra to reach radioactive equilibrium (~ 97%) with its decay products (^{222}Rn, ^{218}Po, ^{214}Pb and ^{214}Bi), as well as ^{228}Ra reached (~ 99 9%) with descendant, ^{228}Ac. The ^{226}Ra and ^{228}Ra activities were determined by γ-ray spectrometry of ^{214}Bi (E_γ = 609.3 keV; I_γ = 46%) and ^{228}Ac (E_γ = 911.1 keV; I_γ = 29%), respectively. The radium retention efficiency of the MnO_2-cellulose cartridges was determined by the activity ratio measurements of the two filters, connected in series, for each collected sample by the following equation [22]:

$$E = 1 - \frac{Mn2}{Mn1}$$

where Mn1 and Mn2 are the Ra activities in the first and second MnO_2-cellulose cartridges, respectively. The retention efficiencies were typically in the order of 80%. The determined activities were corrected for the efficiency of radium retention and the results were reported in Bq L^{-1}.

4. RESULTS

4.1. Produced Water

4.1.1. Soluble and Particulate-Linked Elements

Table 11 shows the results of the produced water samples from the *Pampo* and *Pargo* Platforms. The high salinity of the produced water was verified by the determination of Cl, Na, Ca and Mg in the samples: *Pampo* (3), Cl = 4.2% m/v, Na = 2.3% m/v, Ca = 1.7 g L^{-1}, Mg = 0.56 g L^{-1}; and *Pargo* (6), Cl = 5.8% m/v, Na = 3.0% m/v, Ca = 4.6 g L^{-1}, Mg = 0.27 g L^{-1}. The elements of interest were determined in the soluble fraction of the produced water and in the particulate material, which included the suspended oil present in the samples.

Table 11. Effluent element concentrations at the *Pampo* and *Pargo* platforms in the soluble fraction and particulate material. Values: Soluble in μg L^{-1} and particulate in μg L^{-1} of filtered effluent

Element		Effluent samples		
		Pampo (3)	*Pampo* (4)	*Pargo* (6)
As	soluble particulate	0.17 ± 0.07 < 0,02	0.31 ± 0.06 < 0,02	0.12 ± 0.09 *n.m.*
Ba	soluble particulate	(5.58 ± 0.11) x 10^3 6.25 ± 1.05	(7.34 ± 0.10) x 10^3 10.88 ± 2.13	(8.93 ± 0.08) x 10^3 *n.m.*
Cd	soluble particulate	0.55 ± 0.28 0.02 ± 0.005	0.77 ± 0.18 0.02 ± 0.005	1.22 ± 0.63 *n.m.*
Co	soluble particulate	0.34 ± 0.05 0.09 ± 0.02	0.14 ± 0.05 0.03 ± 0.009	0.39 ± 0.15 *n.m.*
Cu	soluble particulate	36.0 ± 2.16 4.11 ± 1.23	37.0 ± 1.85 1.20 ± 0.31	27.0 ± 3.24 *n.m.*
Fe	soluble particulate	23.0 ± 5.01 66.81 ± 11.43	< 7 108.24 ± 18.67	1.24 x 10^3 *n.m.*
Mo	soluble particulate	< 0.01 < 0.04	1.14 < 0.04	0.05 *n.m.*
Ni	soluble particulate	8.64 ± 4.87 5.91 ± 0.66	4.85 ± 1.29 0.55 ± 0.08	12.14 ± 2.42 *n.m.*
Pb	soluble particulate	5.16 ± 1.27 0.26 ± 0.06	4.04 ± 0.93 0.29 ± 0.06	13.56 ± 8.00 *n.m.*
V	soluble particulate	0.46 ± 0.24 1.43 ± 0.34	0.20 ± 0.06 0.55 ± 0.08	0.15 ± 0.08 *n.m.*
Zn	soluble particulate	35.0 ± 1.22 6.86 ± 1.34	< 8 3.22 ± 0.95	127.0 ± 6.35 *n.m.*

n.m. not measured.

Table 12. Ba, ^{226}Ra, ^{228}Ra and ^{210}Pb concentrations in produced water from the *Bacia de Campos* area

Platform (sample)	Ba (mg/L)	^{226}Ra (Bq L^{-1})		^{228}Ra (Bq L^{-1})		^{210}Pb (Bq L^{-1})	
		soluble	particulate	soluble	particulate	soluble	particulate
Namorado 1	2.7 ± 0.5	0.016 ± 0.006	*n.m.*	<0.05	*n.m.*	0.061 ± 0.013	*n.m.*
Pargo (1)	25.7 ± 0.4	6.0 ± 0.7	*n.m.*	8.2 ± 0.4	*n.m.*	0.208 ± 0.048	*n.m.*
Pampo (1)	*n.m.*	2.28 ± 0.15	*n.m.*	1.60 ± 0.34	*n.m.*	1.936 ± 0.084	*n.m.*
Pargo (2)	*n.m.*	2.52 ± 0.16	*n.m.*	12.0 ± 1.1	*n.m.*	0.310 ± 0.023	*n.m.*
Namorado 1	1.4 ± 0.2	0.059 ± 0.022	1.11 ± 0.10	0.190 ± 0.053	< 0.10	*n.m.*	0.444 ± 0.040
Namorado 2	2.1 ± 0.4	0.076 ± 0.021	1.73 ± 0.11	0.100 ± 0.046	4.28 ± 0.44	*n.m.*	0.573 ± 0.025
Garoupa 1	15.8 ± 0.3	3.71 ± 0.25	0.26 ± 0.03	8.11 ± 0.87	0.24 ± 0.06	*n.m.*	0.052 ± 0.010
Cherne 1	8.0 ± 0.1	2.74 ± 0.20	0.59 ± 0.04	1.24 ± 0.41	0.43 ± 0.10	*n.m.*	0.059 ± 0.010
Cherne 2	3.81 ± 0.07	0.088 ± 0.025	3.02 ± 0.18	0.350 ± 0.065	6.08 ± 0.65	*n.m.*	0.410 ± 0.026
Pampo (2)	13.0 ± 0.3	1.95 ± 0.15	1.3 ± 0.1	0.73 ± 0.31	5.2 ± 0.6	0.107 ± 0.029	0.073 ± 0.016
Pargo (3)	13.0 ± 0.3	2.50 ± 0.18	2.9 ± 0.3	4.29 ± 0.53	5.1 ± 0.5	0.086 ± 0.031	0.163 ± 0.024
Pargo (4)	16.7 ± 0.2	4.80 ± 0.35	6.24 ± 0.36	4.62 ± 0.82	8.6 ± 1.1	0.151 ± 0.062	0.561 ± 0.065
Pargo (5)	15.7 ± 0.4	3.91 ± 0.30	4.88 ± 0.28	3.81 ± 0.70	7.32 ± 0.80	< 0.06	0.150 ± 0.021
Pampo (3)	5.6 ± 0.1	0.82 ± 0.06	0.023 ± 0.005	1.84 ± 0.28	< 0.044	0.39 ± 0.03	0.040 ± 0.008
Pampo (4)	7.3 ± 0.1	1.90 ± 0.13	4.2 ± 0.5	2.3 ± 0.35	9.6 ± 0.7	< 0.11	0.05 ± 0.01
Pargo (6)	8.93 ± 0.08	1.60 ± 0.16	4.15 ± 0.24	2.71 ± 0.40	9.65 ± 0.98	< 0.10	0.211 ± 0.030
Namorado 1	0.36 ± 0.03	0.011 ± 0.007	0.400 ± 0.028	<0.05	1.11 ± 0.11	0.040 ± 0.011	0.028 ± 0.010
Cherne 2	0.78 ± 0.02	0.183 ± 0.047	1.263 ± 0.077	<0.18	2.89 ± 0.30	< 0.063	0.064 ± 0.011

n.m.: not measured.

Despite the great variability of metal concentrations in the produced water found in the literature [4], the values observed in the present study were compatible. The barium concentrations were lower than the effluent concentrations reported by Roe Utvik [23]: Brage (228 mg L^{-1}), Oseberg F (107 mg L^{-1}), Oseberg C (142 mg L^{-1}) and Troll (147 mg L^{-1}) platforms. For As, Mo, V and U, was observed lower concentrations than those normally found in seawater. Co and Cd concentrations were 18-40 and 4-14 times higher than the concentrations found in seawater, respectively. Barium concentrations in the produced water were 1000 times higher than in seawater. Cu, Ni and Pb concentrations were approximately 50, 18 and 170 times higher than in seawater, respectively. No major differences were observed between the concentrations of produced water in both of the studied platforms, a difference that was not possible to estimate due to the lack of a statistically appropriate number of samples.

According to Fonseca [24], the range of metal concentrations in the *Bacia de Campos* oil is: V: 9.8 – 45.0 mg kg^{-1}; Ni: 7.3 – 33.6 mg kg^{-1}; Fe: 3.66 – 182.2 mg kg^{-1}; Cu: 0.08 – 13.94 mg kg^{-1}; Zn: 0.73 – 154.2; Pb: 0.05 – 2.75 mg kg^{-1}. Consequently, the metal content in the effluent samples, especially nickel and vanadium which are porphyrin components present in oil, may vary depending on the oil content dispersed in the produced water. There were significant amounts of barium, iron, nickel and zinc in the particulate material of the studied samples. Vanadium, as expected, was present in higher concentrations in the particulate material when compared to the soluble fraction. The metal concentrations in the effluent were below the maximum permissible values established by the National Council for the Environment (CONAMA) for effluents [25], with the exception of Ba, which was present in concentrations above 5 mg L^{-1}.

4.1.2. Radionuclides. Correlation between Ra Isotopes and Ba

To characterize the natural radionuclides and the correlation of radium isotopes with barium concentrations in the produced water from the *Bacia de Campos* area, three other fixed oil production platforms were studied. Table 12 presents the results of the Ba, ^{226}Ra, ^{228}Ra and ^{210}Pb concentrations in produced water of the five platforms.

The activity concentrations measured in the soluble fraction were in the range of 0.011 – 6.0 Bq L^{-1} for ^{226}Ra, < 0.05 – 12.0 Bq L^{-1} for ^{228}Ra and < 0.06 – 1.9 Bq L^{-1} for ^{210}Pb and in the particulate material were in the range of 0.023 – 6.2 Bq L^{-1} for ^{226}Ra, 0.04 – 9.7 Bq L^{-1} for ^{228}Ra and 0.028 – 0.57 Bq L^{-1} for ^{210}Pb.

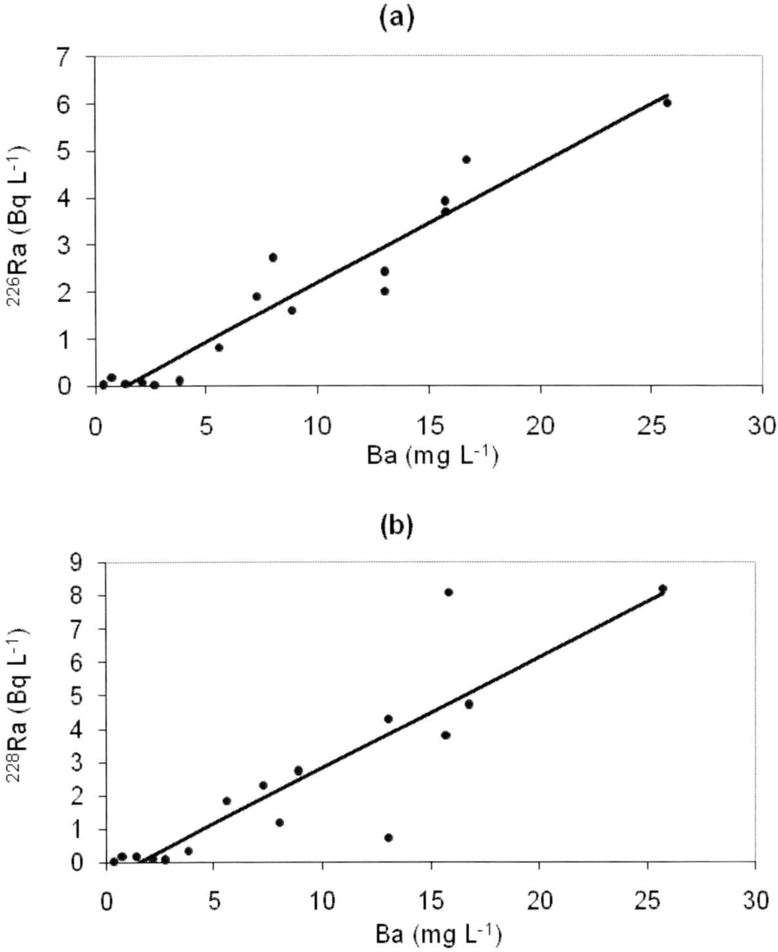

Figure 7. Correlation between the Ba and ^{226}Ra (a) and ^{228}Ra (b) concentrations in produced water at the *Bacia de Campos* area.

The results obtained for ^{226}Ra and ^{228}Ra showed similar values to those reported for produced water from oil and gas platforms in the North Sea [3, 23], Mississippi, USA [26], and other produced waters in USA, Russia and Germany [27].

Ba concentrations were in the range of 0.36 - 25.7 mg L^{-1}. These concentrations are in the range of values reported for produced water from the Snorre, Ekofisk Tor and Ekofisk Kilo fields, and are 4-40 times lower than those reported by Røe [23] for produced water from the Oseberg, Brage and Troll of the North Sea fields.

The results show a strong correlation between Ba and radium isotopes (^{226}Ra: R = 0.926; ^{228}Ra: R = 0.785) in produced water (see figure 7). For the *Pampo* and *Pargo* platforms, the correlation was R = 0.897 for ^{226}Ra and R = 0.737 for ^{228}Ra. The lower correlation observed for ^{228}Ra can be attributed, in part, due to the choice of the radiochemical method. The limit of detection and associated error were compromised by the presence of high contents of ^{226}Ra, as noted by Godoy et al. [7]. The strong correlation can be attributed to similar geological characteristics of the studied area.

4.2. Seawater

4.2.1. Trace Elements in Soluble Fraction

Tables 13 and 14 show the element concentrations in seawater samples (soluble fraction) collected around the *Pampo* and *Pargo* platforms at the *Bacia de Campos* area. Concentrations of dissolved arsenic around the two platforms were very uniform (*Pampo*: 1.36 ± 0.06 μg/L; *Pargo*: 1.32 ± 0.12 μg L^{-1}). Thus, there were no significant differences between the sampling sites and between the three studied depths (ANOVA, p = 0.95). The coefficient of variation was 4.3% for *Pampo* and 9.4% for *Pargo*. Considering all data as belonging to the same population, the coefficient of variation was of only 7.3%. The obtained arsenic values are in the range of the values determined by Pozebon et al. [28]: 1.13 - 1.85 μg L^{-1} in seawater samples. Based on these results and those obtained for the produced water, it can be stated that the arsenic is not a critical element for the purpose of environmental monitoring related to the studied two platforms.

The mean concentration of soluble barium around the platforms was of 7.0 ± 0.8 μg L^{-1}, which is within the range of concentrations reported for unpolluted seawater, of 5-10 μg L^{-1} [29]. Applying the ANOVA test (p = 0.95) for both platform data sets, it was observed that both sets belong to the same population. Tests were also conducted to observe the variation of barium concentrations at the different studied depths, showing no significant differences. At the *Pampo* platform, for points 4C and 4D, it was observed the higher values of barium. However taking into account the similarity of the concentrations from the two platforms, where the coefficient of variation was 11 %, these values maybe considered as normal once this fact can be explained by the expected variability for environmental data.

Table 13. Concentration of metals and elements of interest in seawater samples (soluble fraction) in the vicinity of the *Pampo* platform. Values in μg L^{-1}

Sample	Platform Distance (m)	Level	Depth (m)	As	Ba	Cd	Co	Cu	Mn	Mo	Ni	Pb	V	U
4A	500	Surface	5	1.33 ± 0.08	6.63 ± 0.54	0.012 ± 0.007	0.028 ± 0.003	2.76 ± 0.87	< 0.01	13.27 ± 0.88	0.32 ± 0.13	1.49 ± 0.08	1.48 ± 0.14	3.19 ± 0.34
		Middle	45	1.40 ± 0.09	6.12 ± 0.15	0.031 ± 0.007	0.028 ± 0.001	2.00 ± 0.56	< 0.01	14.12 ± 0.85	0.43 ± 0.07	0.36 ± 0.11	1.45 ± 0.09	3.09 ± 0.26
		Bottom	99	1.42 ± 0.11	6.46 ± 0.17	0.017 ± 0.006	0.029 ± 0.001	1.27 ± 0.09	9.96 ± 0.49	13.86 ± 2.63	0.51 ± 0.05	0.26 ± 0.03	1.33 ± 0.09	2.93 ± 0.25
4B	500	Surface	5	1.36 ± 0.08	6.29 ± 0.20	0.012 ± 0.005	0.032 ± 0.003	0.72 ± 0.10	< 0.01	24.26 ± 0.86	0.61 ± 0.06	< 0.03	1.53 ± 0.16	2.93 ± 0.03
		Middle	50	1.33 ± 0.06	6.46 ± 0.32	0.006 ± 0.003	0.032 ± 0.006	0.60 ± 0.08	0.61 ± 0.24	14.74 ± 1.34	0.49 ± 0.08	< 0.03	1.72 ± 0.08	3.10 ± 0.58
		Bottom	102	1.42 ± 0.06	6.80 ± 0.32	0.013 ± 0.006	0.030 ± 0.005	0.37 ± 0.02	< 0.01	14.35 ± 1.19	0.29 ± 0.07	< 0.03	1.46 ± 0.15	2.96 ± 0.04
4C	500	Surface	5	1.31 ± 0.07	8.67 ± 0.68	0.009 ± 0.002	0.029 ± 0.004	0.22 ± 0.01	< 0.01	13.64 ± 1.55	0.44± 0.08	< 0.03	1.55 ± 0.15	3.01 ± 0.10
		Middle	45	1.34 ± 0.09	8.50 ± 0.36	0.015 ± 0.005	0.030 ± 0.003	0.19 ± 0.03	< 0.01	13.92 ± 0.75	0.21 ± 0.10	< 0.03	1.32 ± 0.15	2.94 ± 0.49
		Bottom	97	1.41 ± 0.06	8.84 ± 0.85	0.064 ± 0.010	0.032 ± 0.006	0.51 ± 0.07	0.26 ± 0.11	13.03 ± 1.27	0.60 ± 0.07	< 0.03	1.50 ± 0.11	2.80 ± 0.33
4D	500	Surface	5	1.29 ± 0.08	7.65 ± 0.15	0.017 ± 0.006	0.028 ± 0.005	0.21 ± 0.10	0.04 ± 0.01	12.42 ± 1.27	0.73 ± 0.08	< 0.03	1.65 ± 0.10	2.89 ± 0.10
		Middle	50	1.34 ± 0.10	7.14 ± 0.27	0.017 ± 0.008	0.029 ± 0.003	< 0.07	< 0.01	13.73 ± 0.77	0.34 ± 0.11	< 0.03	1.35 ± 0.12	2.86 ± 0.49
		Bottom	102	1.45 ± 0.07	7.82 ± 0.22	0.037 ± 0.003	0.032 ± 0.002	0.50 ± 0.15	0.50 ± 0.22	13.56 ± 2.16	0.67 ± 0.06	< 0.03	1.61 ± 0.11	2.54 ± 0.29
5	250	Surface	5	1.26 ± 0.09	5.95 ± 0.30	0.010 ± 0.001	0.030 ± 0.004	< 0.07	0.23 ± 0.09	13.67 ± 1.31	0.62 ± 0.07	< 0.03	1.57 ± 0.10	2.84 ± 0.22
		Middle	50	1.45 ± 0.06	6.46 ± 0.27	0.011 ± 0.001	0.027 ± 0.005	0.15 ± 0.03	< 0.01	12.94 ± 1.08	0.56 ± 0.05	< 0.03	1.46 ± 0.10	2.89 ± 0.06
		Bottom	106	1.33 ± 0.05	6.29 ± 0.32	0.064 ± 0.009	0.033 ± 0.002	0.26 ± 0.04	0.17 ± 0.03	15.72 ± 0.79	0.67 ± 0.09	< 0.03	1.48 ± 0.13	3.24 ± 0.42
6	1000	Middle	50	1.31 ± 0.07	6.29 ± 0.15	0.120 ± 0.014	0.031 ± 0.001	0.36 ± 0.09	0.76 ± 0.40	14.14 ± 0.87	0.81 ± 0.12	< 0.03	1.73 ± 0.07	2.89 ± 0.08
Mean				1.36	7.02	0.028	0.030	0.68	0.37	14.46	0.52		1.51	2.94
SD				0.06	0.96	0.030	0.002	0.77	0.24	2.72	0.17		0.12	0.16

Table 14. Concentration of metals and elements of interest in seawater samples (soluble fraction) in the vicinity of the *Pargo* platform. Values in μg L^{-1}

Sample	Platform distance (m)	Level	Depth (m)	As	Ba	Cd	Co	Cu	Mn	Mo	Ni	Pb	V	U
PE	0	Surface	3	*n.m.*	*n.m.*	0.014 ± 0.006	0.046 ± 0.006	57.86 ± 0.39	4.50 ± 0.04	*n.m.*	0.95 ± 0.36	2.45 ± 0.43	1.56 ± 0.12	*n.m.*
		Middle	43	*n.m.*	*n.m.*	0.011 ± 0.007	0.021 ± 0.003	3.53 ± 0.02	4.53 ± 0.09	*n.m.*	0.23 ± 0.08	0.37 ± 0.02	0.31 ± 0.04	*n.m.*
		Bottom	85	*n.m.*	*n.m.*	0.073 ± 0.010	0.061 ± 0.004	8.39 ± 0.16	28.17 ± 0.73	*n.m*	1.63 ± 0.39	4.76 ± 0.90	0.99 ± 0.06	*n.m.*
4A	500	Surface	5	1.34 ± 0.07	6.80 ± 0.49	0.096 ± 0.015	0.053 ± 0.002	0.78 ± 0.04	9.14 ± 0.51	14.29 ± 0.81	1.15 ± 0.28	0.38 ± 0.06	1.63 ± 0.02	3.69 ± 0.15
		Middle	45	1.45 ± 0.08	6.46 ± 0.19	0.026 ± 0.003	0.028 ± 0.010	0.31 ± 0.09	2.02 ± 0.11	23.52 ± 1.88	0.44 ± 0.11	< 0.03	1.49 ± 0.04	3.65 ± 0.25
		Bottom	97	1.48 ± 0.06	6.29 ± 0.19	0.053 ± 0.021	0.034 ± 0.014	0.49 ± 0.01	0.48 ± 0.02	13.97 ± 2.74	0.74 ± 0.12	< 0.03	1.74 ± 0.10	3.29 ± 0.59
4B	500	Surface	5	1.12 ± 0.10	6.97 ± 0.17	0.026 ± 0.008	0.028 ± 0.007	0.33 ± 0.12	0.19 ± 0.01	44.90 ±4.52	0.28 ± 0.08	< 0.03	1.37 ± 0.06	3.27 ± 0.34
		Middle	45	1.25 ± 0.07	7.14 ± 0.17	0.019 ± 0.008	0.034 ± 0.003	0.22 ± 0.10	0.03 ± 0.01	13.85 ± 0.96	0.39 ± 0.07	< 0.03	1.50 ± 0.06	3.48 ± 0.08
		Bottom	94	1.51 ± 0.05	7.31 ± 0.58	0.045 ± 0.010	0.032 ± 0.006	0.57 ± 0.08	2.36 ± 0.07	13.71 ± 0.90	0.64 ± 0.10	< 0.03	1.71 ± 0.13	3.16 ± 0.20
4C	500	Surface	5	1.32 ± 0.09	7.14 ± 0.29	0.005 ± 0.002	0.027 ± 0.002	0.16 ± 0.02	0.63 ± 0.04	13.96 ± 1.75	0.28 ± 0.06	< 0.03	1.68 ± 0.03	3.31 ± 0.70
		Middle	45	1.37 ± 0.07	6.63 ± 0.22	0.019 ± 0.004	0.030 ± 0.003	<0.07	0.11 ± 0.07	13.41 ± 0.53	0.54 ± 0.04	< 0.03	1.56 ± 0.10	3.01 ± 0.25
		Bottom	98	1.21 ± 0.11	6.80 ± 0.42	0.024 ± 0.002	0.025 ± 0.001	<0.07	0.03 ± 0.01	13.24 ± 1.64	0.33 ± 0.09	< 0.03	1.51 ± 0.09	3.51 ± 0.31
4D	500	Surface	5	1.16 ± 0.09	9.52 ± 0.12	0.012 ± 0.007	0.033 ± 0.001	<0.07	0.06 ± 0.01	13.79 ± 1.82	0.53 ± 0.11	< 0.03	1.42 ± 0.23	3.11 ± 0.46
		Middle	45	1.27 ± 0.08	5.95 ± 0.49	0.026 ± 0.009	0.033 ± 0.001	0.57 ± 0.03	0.38 ± 0.01	13.39 ± 2.24	0.50 ± 0.08	< 0.03	1.49 ± 0.12	3.10 ± 0.28
		Bottom	93	1.22 ± 0.10	6.12 ± 0.22	0.080 ± 0.008	0.042 ± 0.001	0.29 ± 0.08	0.39 ± 0.07	11.92 ± 0.92	1.83 ± 0.20	< 0.03	1.51 ± 0.11	2.60 ± 0.34
5	250	Surface	5	1.35 ± 0.08	7.14 ± 0.10	0.008 ± 0.002	0.033 ± 0.006	0.08 ± 0.04	0.11 ± 0.03	14.98 ± 2.22	1.20 ± 0.13	< 0.03	1.42 ± 0.12	2.92 ± 0.11
		Middle	45	1.49 ± 0.04	7.14 ± 0.37	0.023 ± 0.013	0.032 ± 0.003	0.15 ± 0.03	0.14 ± 0.02	14.65 ± 1.98	0.41 ± 0.06	< 0.03	1.52 ± 0.12	3.29 ± 0.21
		Bottom	95	1.17 ± 0.07	6.63 ± 0.39	0.021 ± 0.011	0.028 ± 0.001	<0.07	0.13 ± 0.04	11.91 ± 2.22	0.56 ± 0.09	< 0.03	1.60 ± 0.08	2.81 ± 0.08
6	1000	Bottom	98	1.38 ± 0.09	7.65 ± 0.39	0.011 ± 0.004	0.031 ± 0.004	<0.07	<0.01	14.58 ± 1.53	0.28 ± 0.04	< 0.03	1.42 ± 0.07	2.98 ± 0.22
Mean*				1.32	6.98	0.031	0.033	0.36	1.08	16.25	0.63		1.54	3.19
SD				0.12	0.81	0.025	0.007	0.22	2.34	8.06	0.42		0.11	0.30

*Mean not including sample PE.; *n.m.* not measured.

Cadmium concentrations were in the range of 0.006 – 0.12 µg L^{-1} for *Pampo* and 0.005 – 0.096 µg L^{-1} for *Pargo*. Applying the Grubbs statistical test, it was observed that the cadmium values corresponding to samples 6 (middle), 4A (surface) and 4D (bottom) at *Pampo* could be rejected. Likewise, no significant differences were observed between the values of data sets from the two platforms, as well as between the three studied depths (ANOVA, p = 0.95). The wide range of cadmium concentrations around the platforms did not reveal the existence of areas affected by the disposal of produced water into the sea. The samples PE, collected at the base of the platform, showed values in the same order of samples collected at 250 and 500 meters distant from the platforms.

The average concentration of cobalt in the *Pampo* and *Pargo* was of 0.031 ± 0.005 µg L^{-1}, showing good result stability. The concentrations of PE samples also showed similar values. Cobalt concentrations were lower than values reported for seawater of the southern coast of Brazil, of 0.06 – 14 µg L^{-1} [28].

Copper concentrations were in the range of <0.07 – 2.76 µg L^{-1} for *Pampo* and, <0.07 – 0.78 µg L^{-1} for *Pargo*. For *Pampo*, at sampling point 4A a concentration factor of 7 (surface), 5 (middle) and 3 (bottom) times higher than the means was observed, of 0.37 ± 0.18 µg L^{-1} for the rest of the samples. Taking into account the copper concentrations in the produced water and the fact that this sampling point is in the direction of the sea current measured at the time of sampling, these concentrations can be explained by the spreading plume of produced water in this direction, despite the high dilution rate due to local currents and weather effects. At *Pargo*, the PE samples collected at the base of the platform revealed high levels of copper. While the produced water is rapidly dispersed immediately after the discharge, the discarded volumes are high and continuous during oil production operations. Thus, if this fact is causing any effect in the water column, this effect will be restricted to the discharge mixing zone. Discarding the result of the PE samples, the copper concentrations in the samples were in the concentration range of 0.20 – 18.9 µg L^{-1} reported in the literature [28]. Similarly, the copper means in samples from both platforms of 0.38 ± 0.19 µg L^{-1} are in agreement with the concentrations reported by Santosa et al. [30] of 0.1 – 0.4 µg L^{-1} for seawater from the northwest Pacific Ocean.

For manganese, concentrations were obtained in the range of < 0.01 – 9.96 µg L^{-1} for *Pampo* and < 0.01 – 9.14 µg L^{-1} for *Pargo*. For both platforms, the higher amount of manganese in seawater (*Pampo*: 4A (bottom); *Pargo*: 4A (surface)) was one hundred times higher than in the rest of the samples.

The PE samples from the *Pargo* platform showed high levels of this metal at the base of the platform, at the on-site discharge area. Even observing a wide dispersion of the manganese concentrations in the studied seawater samples, the values obtained at the two platforms were several times lower than the values of 7.24 – 3.96 x 10^3 μg L^{-1} generally found in the southern coast of Brazil [28].

Molibdenium is a conservative element in seawater. Comparing the values observed at *Pampo*: 12.4 – 24.5 μg L^{-1} and *Pargo*: 11.9 – 44.9 μg L^{-1} with literature data reported by Santosa et al. [30]: 5.5 – 7.0 μg L^{-1} and Collier A. [31] and Sohrin et al. [32] of around 9.8 μg L^{-1}, it was observed an increase in concentrations. Applying the Grubbs statistical test for a confidence level of 95%, molybdenum values were rejected in samples 4B (surface): 24.3 μg L^{-1} and 5 (bottom): 15.7 μg L^{-1} for *Pampo* and 4B (surface): 44.9 μg L^{-1} and 4A (middle): 23.5 μg L^{-1} for *Pargo*. Discarding the outlier values, it was obtained a mean value of 13.67 ± 0.60 μg L^{-1} for *Pampo* and 13.69 ± 0.90 μg L^{-1} for *Pargo*. These values are in the range of concentrations determined by Pozebon et al. [28] of 8.15 - 9.15 μg L^{-1} in seawater. Also, the ANOVA test (p = 0.95) showed no significant differences between the data sets of the two platforms, as well as between the three studied depths.

For nickel, concentrations were obtained in the range of *Pampo*, 0.21 - 0.81 μg L^{-1} for *Pampo* and 0.28 – 1.83 μg L^{-1} for *Pargo*. The Grubbs test considered sample 4D (bottom): 1.83 μg L^{-1} at *Pargo* as an outlier. The mean concentrations were 0.52 ± 0.17 μg L^{-1} and 0.55 ± 0.28 μg L^{-1} for *Pampo* and *Pargo*, respectively. These values are in the range of 0.21 to 4.44 μg L^{-1} reported for coastal waters of southern Brazil [28] and the upper limit of the concentration range of 0.2 - 0.5 μg L^{-1} for the northwest sea Pacific Ocean [33]. The ANOVA test (p = 0.95) showed that the data sets of the two platforms belong to the same population. The results obtained in the additional PE samples revealed no evidence of contamination, even in the mixing zone of the produced water discharge.

Lead levels could be quantified in only a few samples. In most of the samples, concentrations were below the LOD of the method, of 0.03 μg/L. Concentrations above the LOD were detected at sampling points 4A (surface, middle and bottom) at *Pampo* and 4A (surface) and PE (surface, middle and bottom) at *Pargo*. Even considering the lead concentrations in sampling point A4 at both platforms (500 m distance) as being low, these concentrations can be explained by the fact that these stations are in the direction of the sea current measured at the sampling time and are potentially influenced by the spreading plume of produced water.

Vanadium, as well as molybdenum and uranium, is a conservative element in seawater and its concentration is in the range of 1.0 - 2.0 μg L^{-1} [30, 31]. The vanadium values observed in this study showed constant concentrations, with a total mean value of 1.52 ± 0.12 μg L^{-1}, with only 7.8% relative standard deviation. Like nickel, vanadium has been used as an indicator of environmental impacts arising from discharges or oil spills at sea. Therefore, the constancy of the results obtained in this study rules out the possibility of contamination of the region.

The uranium average calculated for both platforms was of 3.08 ± 0.28 μg L^{-1}, confirming the conservative nature of this element in the water column, as well as confirming that this is not a critical element for the purpose of environmental monitoring related to offshore oil production.

Selenium also proved to be a non-relevant element for environmental monitoring in the region and the sample concentrations were below the method LOD: 0.02 μg L^{-1}.

4.2.2. Elements Associated to Particulate Material

Tables 15 and 16 present the concentrations of the elements associated to particulate material in the seawater: the 4A and 4B samples for *Pampo* and 4A and 5 for *Pargo* platform. The results are given in μg per liter of filtered water. As could not be performed a comparison of the levels found in the particulate expressed directly in particulate mass, it was decided to standardize the studied elements in relation to the mass of aluminum.

The mean values of Al, Cu and Ba in the particulate matter in seawater at the *Pargo* platform were 4, 3 and 2 times higher, respectively, than those obtained at the *Pampo* platform. Likewise, Ni, Cr and Cd values at the *Pampo* platform were 4, 3 and 2 times higher, respectively, than those determined at the *Pargo* platform.

For the rest of the elements, this ratio was of, approximately, 1:1. Applying the Grubbs test, it was observed that the Ba *Pampo* platform value in sample 4A (surface) of 0.045 μg L^{-1}, and the Cu value in sample 4C (surface) of 0.19 μg L^{-1} were rejected. Similarly, for Ni in sample 4B (bottom) of 0.74 μg L^{-1}, the log-normal distribution test showed that this value should be rejected.

Based on those elements considered potentially critical: barium, vanadium and nickel, after performing an ANOVA test (p=0.95) between data populations, no significant differences between the sampling stations for each platform or even between different depths of the sample were observed.

Table 15. Elements associated to the particulate materiel in the seawater samples 4A and 4B collected from the surrounding area of the *Pampo* platform. Values in μg L^{-1}

Element	4A (Surface)	Al ratio	4A (Middle)	Al ratio	4A (Bottom)	Al ratio	4B (Surface)	Al ratio	4B (Middle)	Al ratio	4B (Bottom)	Al ratio
Al	0.44	1.00	0.30	1.00	1.63	1.00	1.64	1.00	0.10	1.00	1.23	1.00
V	0.014	0.032	0.014	0.047	0.032	0.020	0.028	0.017	0.038	0.36	0.031	0.025
Cr	0.090	0.20	0.040	0.14	0.170	0.104	0.16	0.099	< 0.025		1.35	1.10
Fe	1.32	2.96	0.98	3.32	2.60	1.59	1.18	0.72	1.45	14.0	6.20	5.05
Ni	0.063	0.14	0.095	0.32	0.083	0.051	0.052	0.032	0.063	0.61	0.74	0.60
Cu	0.042	0.094	0.025	0.084	0.037	0.023	0.076	0.046	0.077	0.74	0.046	0.038
Zn	< 0.05		< 0.05		0.075	0.05	< 0.050		< 0.050		< 0.050	
As	0.007	0.016	0.004	0.014	0.002	0.001	0.005	0.003	0.004	0.038	0.005	0.004
Se	0.014	0.032	< 0.01		< 0.01		0.051	0.031	0.036	0.35	0.017	0.014
Ag	< 0.003		< 0.003		< 0.003		< 0.003		< 0.003		0.004	0.003
Cd	0.001	0.002	0.002	0.007	0.004	0.002	0.001	0.0006	0.001	0.010	0.0003	0.0002
Sn	< 0.010		< 0.010		< 0.010		< 0.010		< 0.010		< 0.010	
Ba	0.045	0.101	0.01	0.034	0.015	0.009	0.027	0.016	0.023	0.22	0.025	0.020
Pb	< 0.008		< 0.008		0.011	0.007	< 0.008		< 0.008		< 0.008	

Table 16. Elements associated to the particulate material in the seawater samples 4A and 5 collected from the surrounding area of the *Pargo* platform. Values in μg L^{-1}

Element	4A (Surface)	Al ratio	4A (Middle)	Al ratio	4A (Bottom)	Al ratio	5 (Surface)	Al ratio	5 (Middle)	Al ratio	5 (Bottom)	Al ratio
Al	2.35	1.00	0.94	1.00	4.62	1.00	4.57	1.00	1.87	1.00	2.37	1.00
V	0.013	0.006	0.015	0.016	0.019	0.004	0.014	0.003	0.009	0.005	0.012	0.005
Cr	< 0.025	0.050	0.050	0.053	0.057	0.012	0.038	0.008	< 0.025		< 0.025	
Fe	1.03	0.44	1.05	1.12	3.16	0.68	1.36	0.30	0.70	0.37	1.19	0.50
Ni	0.009	0.004	0.020	0.021	0.013	0.003	0.048	0.010	0.031	0.017	0.025	0.011
Cu	0.031	0.013	0.043	0.046	0.32	0.068	0.18	0.040	0.096	0.051	0.14	0.059
Zn	< 0.050		< 0.050		0.14	0.030	0.13	0.028	0.055	0.029	< 0.050	
As	0.002	0.0008	0.005	0.005	0.004	0.0009	0.005	0.001	0.003	0.002	0.004	0.002
Se	0.019	0.008	< 0.010		< 0.010		< 0.010		< 0.010		0.014	0.006
Ag	< 0.003		< 0.003		< 0.003		< 0.003		< 0.003		< 0.003	
Cd	0.001	0.0004	0.0008	0.0008	< 0.0002		0.0003	0.00006	0.002	0.001	< 0.0002	
Sn	< 0.010		< 0.010		< 0.010		< 0.010		< 0.010		< 0.010	
Ba	0.029	0.012	0.021	0.023	0.040	0.009	0.040	0.009	0.060	0.032	0.017	0.007
Pb	< 0.008		< 0.008		0.017	0.004	0.013	0.003	0.025	0.013	0.011	0.005

The mean Ba concentrations associated with particulate material was of 0.022 ± 0.014 μg L^{-1}, which represents a small fraction (0.3%) of this component in seawater.

For Ni and V, the mean concentration was of 0.046 ± 0.024 μg L^{-1} and 0.016 ± 0.012 μg L^{-1}, that represents a fraction of 1% and 8% of Ni and V in seawater, respectively.

4.2.3. Radionuclides in Soluble Fraction

Using 20 L seawater samples, the data obtained for ^{226}Ra, ^{228}Ra and ^{210}Pb were below the limit of detection of the method used (5 mBq L^{-1}). This result is applied for both platforms, in accordance with the levels of these radionuclides found in the literature for coastal waters, of the same order of magnitude, ranging from 1 to 5 mBq L^{-1}. In waters of the continental shelf region, due to a smaller influence of the coastal region, the values reported in the literature are even lower [34, 35].

Radium was additionally determined in six sampling stations around the *Pargo* platform. After pre-concentration of radium isotopes from 400 L samples onto cellulose cartridges coated with MnO_2, concentrations of about 1.5 mBq L^{-1} were measured and which are in agreement with literature values reported for unpolluted seawater [36]. Table 17 presents the results of the activity concentrations of ^{226}Ra and ^{228}Ra in samples of surface water around the *Pargo* platform.

With these results it was not possible to observe any influence of the production platforms in the ^{226}Ra, ^{228}Ra and ^{210}Pb concentrations in the water column, even at a distance of 250 meters from the discharge point of the produced water. A more detailed quantification is possible only by sampling even higher amounts of water and/or elevated counting time.

Table 17. Activity concentrations of ^{226}Ra and ^{228}Ra in seawater samples at the *Pargo* platform. Pre-concentration in MnO_2-cellulose cartridges

Sample	Distance (m)	^{226}Ra (mBq L^{-1})	^{228}Ra (mBq L^{-1})
4A	500	1.73 ± 0.40	< 3.0
4B	500	1.30 ± 0.37	< 1.50
4C	500	< 1.35	< 2.81
4D	500	< 1.35	< 3.0
5	250	< 1.27	< 2.96
6	1000	0.95 ± 0.29	0.75 ± 0.45

4.2.4. Radionuclides Associated to Particulate Material

Table 18 shows the results obtained for the radionuclides associated with particulate matter in samples from the *Pampo* and *Pargo* platforms.

Table 18. Activity concentrations of ^{226}Ra, ^{228}Ra and ^{210}Pb in particulate material samples at the *Pampo* and *Pargo* platforms

Sample	Distance (m)	Level	Depth (m)	^{226}Ra (mBq L^{-1})	^{228}Ra (mBq L^{-1})	^{210}Pb (mBq L^{-1})
Pampo platform						
4A	500	Surface	5	0.21	< 0.66	12.6
		Middle	45	0.14	< 0.83	1.6
		Bottom	99	0.16	0.61	< 0.84
4B	500	Surface	5	0.23	< 0.65	3.5
		Middle	50	< 0.17	0.92	0.79
		Bottom	102	0.45	< 0.67	0.66
4C	500	Surface	5	< 0.16	0.36	6.3
		Middle	45	< 0.10	0.88	8.0
		Bottom	97	< 0.39	< 0.34	< 0.44
4D	500	Surface	5	< 0.29	1.20	2.6
		Middle	50	0.15	< 0.44	1.2
		Bottom	102	< 0.10	0.26	4.6
5	250	Surface	5	0.70	< 0.73	3.2
		Middle	50	< 0.26	0.47	3.0
		Bottom	106	0.22	< 0.37	< 0.81
6	1000	Bottom	50	0.20	< 0.53	7.4
Pargo platform						
A4	500	Surface	5	< 0.18	0.57	2.9
		Middle	45	< 0.36	< 0.33	2.2
		Bottom	97	< 0.2	1.40	3.5
5	250	Surface	5	< 0.42	< 0.39	3.3
		Middle	45	< 0.17	0.60	1.2
		Bottom	95	< 0.23	<0.48	1.5

The concentration of radionuclides in the particulate material is certainly dependent on the amount of the particulate material matter itself. In the study area a low turbidity in seawater was observed, and the activities of ^{226}Ra, ^{228}Ra and ^{210}Pb were, like those in solution, very low, often close to the limit of detection of the method. The radium emitted in its dissolved form should remain in solution, while ^{210}Pb is associated with the particulate matter.

Radium, generated by the produced water and discharged into the sea by effluents in the form of co-precipitated barium sulfate, could not be detected in the suspended particle fraction collected from the vicinity of the studied platforms. Likewise, higher concentrations of ^{210}Pb than ^{226}Ra and ^{228}Ra in the particulate matter are expected, due to the contribution of ^{210}Pb of atmospheric origin.

4.3. Degree of Dilution

Table 19 presents the degree of dilution of the metals and radionuclides in seawater with respect to the effluents of the studied platforms. The results showed a high degree of dilution for barium and the radium isotopes. Also, there was a significant degree of dilution for copper, cadmium, nickel and ^{210}Pb. Arsenic, molybdenum and vanadium in the effluents showed concentrations below those found in seawater.

Table 19. Degree of dilution of metals and radionuclides in seawater with respect to the produced water samples from the *Pampo* and *Pargo* platforms

Element	Effluent samples – Platforms (sampling number)		
	Pampo (3)	*Pampo* (4)	*Pargo* (6)
Ba	7.97 x 10^3	1.05 x 10^4	1.28 x 10^4
Cd	29	41	64
Co	9	4	11
Cu	180	185	135
Ni	29	16	41
^{210}Pb	98	28	25
^{226}Ra	2.73 x 10^3	6.33 x 10^3	5.33 x 10^3
^{228}Ra	1.42 x 10^3	1.77 x 10^3	2.09 x 10^3

Evaluating the environmental impact of the offshore oil production oceanic platforms in the *Bacia de Campos* area, the concentrations of elements in seawater are similar to those established in the literature for non-impacted regions and are below the limits established by CONAMA for saline water [25].

CONCLUSION

The concentrations of Ba, Cd, Cu, Fe, Mo, Ni, Pb and Zn in the effluents of the platforms were consistent with those reported in the literature and, in many cases, even lower. Concentrations were below the maximum permissible values established by CONAMA for effluents, except for Ba, with concentrations above 5 mg L^{-1}.

The results obtained for ^{226}Ra, ^{228}Ra and ^{210}Pb in effluents showed activity concentrations similar to those reported for produced water from oil and gas platforms in different regions of the world. A strong correlation was observed between Ba and the radium isotopes: ^{226}Ra: R = 0.926 and ^{228}Ra: R = 0.785 in produced water of five platforms in the *Bacia de Campos* area, indicating a similarity of characteristics of the produced water in oil wells of this region.

The highest concentrations of Cu, Mn and Pb were detected at the *Pargo* platform in a sampling point at the foot of the platform at the three studied depths. Significant concentrations of Cu and Pb were detected at the *Pargo* platform in the three depth levels of sampling station 4A (500 m distant), which may be explained by the spreading plume of produced water in this direction.

The concentrations of elements in seawater were similar at both platforms and are in agreement with literature data for non-impacted regions. The values of ^{226}Ra, ^{228}Ra and ^{210}Pb in the seawater were below the limit of detection of the method, of 5 mBq L^{-1}. At the *Pargo* platform, the pre-concentration method of radium isotopes in MnO_2-cellulose cartridges showed values around 1.5 mBq L^{-1}. Results were consistent with published data for these radionuclides in seawater.

With the results of this research can be concluded that, for the studied region, despite the large volumes of produced water discharged into the sea, the concentrations of elements and radionuclides in seawater are close to natural levels, indicating that the dispersion caused by local currents minimizes environmental impacts.

REFERENCES

[1] Scholten, M. C. Th.; Karman, C. C.; Huwer, S. *Toxicol. Lett.* 2000,112-113, 283-288.

[2] Somerville, H. J. Environmental aspects of offshore operations, Microbiological. *Problems in the offshore oil industry*, International Conference, Aberdeen, Scotland. Jonh Wiley and Sons: London, UK, 1987, 105-124.
[3] The EandP Forum, Oil Industry International Exploration and Production Forum. North Sea Produced water: Fate and Effects in the Marine Environment, Report No. 2.62/204. London, UK, 1994.
[4] Neff, J., Sauer, T. C.; Maciolek, N. Fate and effects of produced water discharges in Nearshore Marine water; API Publication No. 4472, American Petroleum Institute, Washington D. C., 1989.
[5] Underhill, P. T. *Naturally occurring radioactive material: Principle and practices*; CRC press, St. Lucie Press: Delray Beach, FL, 1996.
[6] Heaton, B.; Lambley, J. Appl Radiat Isto. 1995, 46(6-7), 577-581.
[7] Godoy, J. M. Radiatividade natural em atividades petrolíferas: Ocorrência e aplicações. IV Seminário Química Analítica na Petrobrás; PETROBRAS: Rio de Janeiro, RJ, 1996, 253-258.
[8] Carney, R. S. A review of study designs for the detection of long term environmental effects of offshore petroleum activities. In: *Long-term environmental effects of offshore oil and gas development activities*; Elsevier Applied Science: London, UK, 1987, 651-690.
[9] Jerez Veguería S. F.; Godoy J. M.; Miekeley N. *J. Environ. Radioact.* 2002, 62(1), 29-38.
[10] Michel, J., Moore, W.S.; King, P.T. Anal Chem 198, 153, 1885-1889.
[11] Willie, S. N.; Iida, Y.; McLaren, J. W. *J. Anal. At. Spectrom.* 1998, 19(3), 67-72.
[12] Byrne, J. P.; Chapple, G. At Spectrosc. 1998,19(3), 116-120.
[13] Pozebon, D.; Dressler, V. L.; Curtius, A. J. *J. Anal. At. Spectrom.* 1998, 13, 363- 369.
[14] Bloxham, M. J.; Hill, S.J.; Worsfold, P.J. *J. Anal. At. Spectrom.* 1994, 9, 935-938.
[15] Tan, S. H.; Horlick, G. *Appl Spectrosc.* 1986, 40, 445-460.
[16] Reed, N. M.; Cairns, R. O.; Hutton, R. C. *J. Anal. At. Spectrom.* 1994, 9, 881-896.
[17] Story, W. C.; Caruso, J. A.; Heitkemper, D. T.; Perkins, L. *J. Chromatogr. Sci.* 1992, 30, 427-432.
[18] Sheppard, B. S.; Shen, W. L.; Caruso, J. A., Heitkemper, D. T.; Fricke, F. L. *J. Anal. At. Spectrom.* 1990, 5(8), 697.
[19] Jerez Vegueria S. F.; Godoy J. M.; Campos R. C.; Gonçalves R. A. *Microchem J.* 2013, 106, 121-128.

[20] Godoy, J.M.; Godoy, M.L.; Carvalho, Z. L. *J. Radioanal. Nucl. Chem.* 1994, 182(1), 165-169.

[21] Adams, F.; Dams, R. Applied gamma-ray spectrometry. 2nd Ed.; Pergamon Press: Oxford, 1970.

[22] Scott, M. K.; Moran, S. B. *J. Environ. Radioact.* 2001, 54, 163-174.

[23] Roe Utvik, T. I. R. Chemosphere. 1999, 39(15), 2593-2606.

[24] Fonseca, T. C. Caracterização dos petróleos brasileiros pela técnica de ICP-MS para fins de exploração geoquímica. Tese de Doutorado, Pontificia Universidade Católica do Rio de janeiro, 2000.

[25] CONAMA. Resolução CONAMA nº 357, de 17 de março de 2005. Publicado no Diário Oficial da União nº 053, de 18/03/2005, 2005.

[26] Matthews, J.C.; Bogan, C.; Swann, C.T. and Ericksen, R.L. *Radioact Radiochem.* 2000, 11, 55-65.

[27] Lysebo, I.; Birovljev, A.; Strand, T. NORM in oil production, Occupational doses and environmental aspects; Nordic Society for Radiation Protection, 7th Nordic Radioecology Seminar, Reykjavík, 1996.

[28] Pozebon, D.; Dressler, V. L.; Curtius, A. *J. At. Spectrosc.* 1998, 19(3), 80.

[29] Klinkhammer, G.P.; Chan, L.H. Anal Chim Acta. 1990, 232, 323-329.

[30] Santosa, S. J.; Tanaka, S.; Yamanaka, K. Fresenius *J. Anal. Chem.* 1997, 357, 1122-1127.

[31] Collier, R. W. *Limnol Oceanogr.* 1985, 30(6), 1351-1354.

[32] Sohrin, Y.; Issiki, K.; Kuwamoto, T.; Nakayama, *E. Mar. Chem.* 1987, 22, 95-103.

[33] Santosa, S. J.; Tanaka, S.; Yamanaka, K. *Environ. Monit. Assess.* 1997, 44, 515-528.

[34] Pentreath, R. J. *Nucl Instrum Methods Phys Res.* 1984, 223, 493-501.

[35] Carvalho, F. P. *Sci. Total Environ.* 1997, 196, 151-161.

[36] Elsinger, R.E.; King, P.T.; Moore, W.S. *Anal. Chim. Acta.* 1982, 144, 277-281.

In: Seawater
Editor: Mathew R. White
ISBN: 978-1-62808-606-5

Chapter 3

SELF-PURIFICATION OF SEAWATER: A MEASURE FOR ENVIRONMENTAL REGULATION

Sergei B. Gulin*[*] *and Viktor N. Egorov
The A.O. Kovalevsky Institute of Biology of the Southern Seas (IBSS)
National Academy of Sciences of Ukraine,
Autonomous Republic of Crimea, Ukraine

ABSTRACT

The chapter deals with the development of radiochemoecological approach for evaluation of the maximum allowable input (MAI) of radioactive and toxic substances to marine environments, which combines assessment the capability of seawater for self-purification and toxicity of the pollutants. The net rate of self-purification is proposed to estimate using the time-series data on concentration in seawater of the fallout radionuclides, both well-soluble (^{137}Cs, ^{90}Sr) and particle-reactive ($^{239+240}Pu$). This provides opportunity for direct evaluation of the levels of maximum allowable discharge into marine environments of nuclear and non-nuclear pollutants; for differentiation between their effective and ecological half-lives; for assessment of capacity of aquatic ecosystems in respect of accumulation of contaminants. Quantification of toxicity may be expressed, in general, as the classic values of the maximum allowable

[*] E-mail: s.gulin@ibss.org.ua.

concentrations (MAC), but there is an uncertainty regarding the applicability of many of national and international levels of MAC for environmental regulation of anthropogenic impact in marine ecosystems. This is particularly important for comparative assessment of the influence of nuclear and non-nuclear contaminants having different nature of biological effects and different dimensions. An approach for such the comparison is the evaluation of dosimetric equivalents (e.g., Grey-equivalent), which are used to compare the effects of ionizing radiation and other stressors in aquatic ecosystems. But this would require every time to carry out the parallel ecotoxicological and radioecological dose-effect experiments that is not always accessible. In order to overcome this difficulty, a dimensionless value expressed as a fraction of maximum ability of marine organisms for resistance to radiation or toxic exposures is discussed in this chapter based upon the early reported experiments on the effects of mercury and incorporated radionuclides on the marine unicellular algae. The developed MAI criterion may be useful tool for scientific justification of environmental control of pollution of seawater with the nuclear and non-nuclear substances at a common methodological basis. The chapter highlights also the questions related to environmental assessment of the state of marine ecosystems, which have been subjected to the structural changes due to anthropogenic influence, and their response through biogeochemical processes of compensatory homeostasis.

INTRODUCTION

It is generally difficult to evaluate the relative importance of numerous stress factors, both natural and anthropogenic, on the indicators of marine ecosystems health. Many of these factors act interactively so that, for example, the overall effect of changes in the inputs of nutrients, organic wastes and toxic metals would be difficult to predict with the current understanding of the elements turnover, ecosystem dynamics and transport pathways of contaminants. It is an even greater challenge to predict the time scale over which the changes of aquatic ecosystems is likely to occur, especially since environmental policy may also change over time and may vary from one country to another. Because of limitations in the current ability to describe the integrated effects of different processes, it is not at all clear how the state of aquatic ecosystems may change, even though the history of environmental disturbance over the last years is reasonably well documented (e.g., Polikarpov. Egorov, 1986; GESAMP, 1982, 1990; Muller et al., 2001). This problem is strengthen by a lack of general-purpose indexes for comparable assessments of environmental impacts and risks in marine ecosystems, which considerably differ from each other and are influenced with a large number of many-kind harmful factors, not only water pollution and eutrophication, but also another types of anthropogenic activities directly resulting in changes of run-off, salinity, transparency of water, circulation pattern, sedimentation regime, loss of biodiversity, decline of habitats of biota species etc.

Thus, in order to gain a clearer understanding of the processes presently influencing the marine ecosystems, it is necessary to synthesize all data available on negative (or positive) changes in aquatic environments, to quantify the anthropogenic pressure, to identify the time scale of the changes and to differentiate the human influence from those induced by climatic perturbations or by the natural cycles of biologically active chemicals. Such an approach requires assessment of capability of marine ecosystems to recover naturally, to verify existing models of the self-purification capacity of marine basins and to identify reversible and irreversible consequences of anthropogenic impacts on the marine life and its environment (e.g., GESAMP, 1990).

Assessment of environmental state of marine ecosystems is based traditionally upon the use of maximum allowable concentrations (MAC) of toxic substances in seawater and biota or threshold limit values (TLV) of pollutants as the measures of anthropogenic pressure (e.g., GESAMP, 1990, and refs therein). These terms differ in their respective meanings: the MAC is a ceiling concentration whereas the TLV is a time-weighted average of the concentration of the hazardous agent in the environment (e.g., Tsuchiya, Harashima, 1965; Polikarpov, 1966). At the same time, the MAC and TLV values are static and they can not be applied directly for estimation of permissible level of the pollutant input to the water basins, because they do not take into account the removal of harmful substances, their biogeochemical transformation, decay or degradation, immobilization and deposition in seabed sediments. Moreover, the values attributable to the MAC or TLV vary depending on the selected criteria for assessment of environmental hazard and ecological response of marine ecosystems.

Other criteria used for environmental assessments of marine ecosystems, such as the ETLs (emission toxicity limits), PEC (predicted environmental concentration), PNEC (predicted no-effect-concentration), RQ (risk quotient) etc., which were developed by the European Commission in the Technical Guidance Document for risk assessment of chemicals (EC, 1996) and in the Water Framework Directive (Dir. 2000/60/EC), are widely applied as qualitative, semi-quantitative or quantitative measures to assess the water quality against pollution, to evaluate ecotoxicity of the discharged wastewater, to estimate the likelihood or degree of impact on ecosystems, and to describe those ecosystems or their compartments that may be in danger as a result of pollution with harmful substances. However, these indexes are mostly used for evaluation of hazard of the local effluents of wastewater, and less applicable for assessment of threats of human impact on the large water bodies (e.g., Polikarpov, Egorov, 1986). Also, the methods for assessment the risk may differ between industries and it pertains to general financial decisions or environmental, ecological, or public health (e.g., GESAMP, 1982, 1990). Finally, the traditional approaches for environmental risk assessment do not consider, frequently, the extent to which damaged habitats are able to recover once the polluting inputs have ceased, while it is essentially necessary to predict: whether the pre-impact ecosystem will eventually be reestablished (although this is the simplest expectation, since ecosystems are highly dynamic, often with several possible stable states, the recovery need not follow the same sequence or time scale, and the system will not therefore necessarily revert to its previous state and structure). Overall, many of the pollution control strategies had been prompted, often post hoc, by evident effects of, for example, excessive discharges of toxic substances, but the current approaches should try to forestall effects by defining potentially hazardous substances and stringently limiting their disposal.

One of the most promising ways to achieve this goal is the estimation of environmental (or receiving, absorptive or assimilative) capacity of aquatic ecosystems, which is defined as their ability to take up an impact without unacceptable changes (GESAMP, 1982, 1990; Izrael, Tsiban, 1983; Polikarpov, Egorov, 1986). But though the capacity is quantifiable for many of particular environments, it is often not easily to define such the unacceptable limits of the anthropogenic pressure on the basis of diverse and variable physical, chemical and biological characteristics of marine ecosystems. Polikarpov and Egorov (1981, 1986) have therefore developed a practicable concept of ‘ecological capacity’ regarding the pollution, which is dynamic parameter showing the balance between pollutant inflow to marine basins and their capability for self-purification. This value is calculated by the quantification and modeling of the pollutant fluxes between the main compartments of marine ecosystems including their deposition in the seabed sediments. At the same time, the ecological capacity does not take into account directly the toxicity of pollutants, while it is clear that even completely poisoned water body can be able for a self-cleaning due to, at least, the pollutant dilution, their lithogenic sedimentation and other abiotic factors.

Therefore, the main purpose of this chapter was to overview the existing models and to further develop radiochemoecological approach for evaluation the maximum allowable input of pollutants to marine basins to be used as a numeric measure for environmental regulation of the human impact, which would combine the rates of self-purification of seawater against the nuclear and non-nuclear pollutants and their toxicity. It was also taken into account, that capability of marine environments for self-cleaning is based frequently upon the assumption that it little depends on the intensity of human impact. This situation was determined by Odum (1983) as a 'resistant homeostasis'. But it is known that such the statements not always correct because some processes governing the purification of seawater may depend on the extent of anthropogenic influence (e.g., Egorov, 2012). Therefore an additional goal of this chapter was to highlight the necessity of a wider applying the concept of compensatory homeostasis for assessments the state of marine ecosystems.

OVERVIEW OF APPROACHES AND DEVELOPMENTS

Data sources and Methodological Framework

The chapter is based on the results of our pre- and post-Chernobyl radioecological monitoring carried out in the Black Sea, across the Mediterranean, Atlantic and Antarctic Oceans within the framework of the following national and international projects governed by International Atomic Energy Agency, EU, NATO and European Scientific Foundation: "Application of Radiotracer Techniques to Study the Processes and Pollution in the Black Sea", IAEA CRP/7400/RB; "Marine Environmental Assessment of the Black Sea", IAEA TCP RER/2/002; "The interactions between the river Danube and the north-western Black Sea, EROS-2000", EU EV5V-CT94-0501; "European River-Ocean Systems: Biogeochemical interactions between the Danube River and the north-western Black Sea, EROS-21", EU ENV4-CT96-0286; "Interaction between eutrophication and trace metals in the Black Sea", NATO CRP/972740; "Environment Toxicology Program", ESF SVF/95/13/E; "Mediterranean Mussel Watch", CIESM; "Marine Environmental Assessment of the Mediterranean Sea", IAEA RER7003/9007/01; "Assessment of the Marine Radioactivity in the Area of the Gulf of Aqaba", IAEA JOR70050 (Polikarpov et al., 1991, 1992, 2008; Egorov et al., 1999, 2010; IAEA, 2004; Gulin, Stokozov, 2005; Thébault et al., 2008; Gulin et al., 2012, 2013).

These studies were fulfilled during 41 scientific cruises on research vessels “Academician Kovalevskiy” (USSR), “Mikhail Lomonosov” (USSR), “Academician Vernadsky” (USSR), “Professor Kolesnikov” (USSR), “Professor Vodyanitskiy” (USSR/Ukraine), “Gorizont” (Ukraine), “Palagruža” (Croatia), “Professor Logachev” (Russia), “Academician Keldysh” (Russia), “Seabell” (Israel), “Meteor” (Germany), Maria S. Merian (Germany), on research submarine “Bentos-300” (USSR) and at the Ukrainian Antarctic station “Academician Vernadsky” (former British station “Faraday”).

Smaller-scale time trends of concentration of the man-made radionuclides and chemical pollutants were observed in Sevastopol Bay, in the western part of the Black Sea and in front of the Danube and Dnepr rivers (e.g., Polikarpov et al., 1991, 1992, 2008; Egorov et al., 1999, 2010).

In marine environments and areas which would otherwise be difficult to investigate due to a lack, incompleteness, poor comparability or absence of the previous observations and monitoring data, the contamination history was reconstructed using the methods of nuclear geochronology, which is based upon the tracing of deposition signatures (or records) of pollutants in the seabed sediments dated with the fallout (^{134,137}Cs, $^{238,239+240}$Pu, ^{241}Am) and naturally occurring (^{234}Th/^{238}U, ^{210}Pb/^{226}Ra) radionuclides (Gulin et al., 1997, 2002, 2003). The parallel measurement of sediment-depth profiles of the persistent non-radioactive pollutants has allowed recovery the deposition chronology of their input to the marine environments at an uninterrupted and comparable basis over the last 50-150 years, i.e. throughout the entire nuclear epoch and whole history of pollution of the marine basins with the harmful substances (Gulin et al., 1995).

The concept of ecological capacity of seawater was developed based on the theory and the semi-empirical models of the kinetics of mineral metabolism of marine biota and the processes of pollutant migration across the abiotic compartments of the ecosystems, by the use of data from laboratory and in situ experiments with the toxic substances labeled with different radioactive and stable isotopes, such as ^{3}H, ^{14}C, ^{15}N, ^{32,33}P, ^{35}S, ^{54}Mn, ^{60}Co, ^{65}Zn, ^{90}Sr, ^{106}Ru, ^{131}I, ^{134,137}Cs, ^{144}Ce, ^{203}Hg, ^{234}Th, ^{242}Pu and ^{243}Am (e.g., Polikarpov, Egorov, 1986; Gulin, 2013).

The equi-dosimetric characteristics of the influence of nuclear and non-nuclear pollutants on marine biota were defined during the laboratory experiments for radiation-ecotoxicological study of the effects of ionizing radiation of incorporated radionuclides with different biological significance (^{14}C, ^{106}Ru, ^{144}Ce) and mercury, in small, medium and lethal doses, on the division rates, dimensional characteristics, photopigments composition and photosynthesis of the Black Sea unicellular algae (Gulin, 2013).

Environmental Capacity: Developments and Definitions

Historically, one of the first quantitative definitions of the limits for accumulation of hazardous substances in aquatic environments was proposed by Agre & Korogodin (1960) and their followers (Korogodina et al., 1996), who have developed concept of radiocapacity of water bodies subjected to radioactive contamination. This capacity was defined as maximal amount of radionuclides in the aquatic ecosystem which does not yet violates its trophic functions, i.e. abundance, diversity, biomass and productivity of biota. The total inventory of radionuclide, at the steady-state conditions and assuming negligible ratio between the biomass and the masses of water and sediments, was expressed as:

$$A = C_W S(H + Kh), \tag{1}$$

where C_W is volume activity of radionuclide in water; S – surface square of water and bottom of the basin (taking into account that when its length and width are much larger than depth, both these areas may be considered as almost identical); H – its average water depth; h – thickness of the contaminated uppermost layer of the bottom sediments; K – concentration factor of radionuclide accumulation in the upper sediments ($K = C_S/C_W$, where C_S is volume activity of radionuclide in the upper sediment layer). Accordingly, the critical amount of radionuclide in such the basin may be defined as:

$$A_K = C_K S(H + Kh), \tag{2}$$

where C_K is highest allowable concentration of radionuclide in the water column, which has yet no effect on biota. If to take into account decay of the radionuclide, then the upper limit of its environmentally 'safe' amount in the basin is:

$$A_E = \lambda C_K S(H + Kh), \tag{3}$$

where $\lambda = \ln 2/T_{0.5}$ is decay constant, $T_{0.5}$ – half-life of radionuclide.

Finally, the following value, which determines the fraction of radionuclide amount deposited in the bottom sediments relative to its total inventory in the basin, was defined by Agre & Korogodin (1960) as a 'radiocapacity factor' showing the redistribution of radionuclide due to the processes of its scavenging in the water column:

$$F = \frac{kh}{H + kh}, \tag{4}$$

The equations (2) and (3) reflect the direction for further development of the theory of environmental capacity towards the concept of assimilative capacity. This capacity refers to the ability of marine ecosystems to receive pollutants without deleterious effects and without damage to the environment, aquatic life or humans who consume the water resources (e.g., Izrael, Tsiban, 1983; Tsiban, Izrael, 1988/1989; Bronfman, 1992; Pearce, Turner, 1990; Leandri, 2009). Quantitatively, the assimilative capacity equals the amount of pollutants that a water body may absorb while maintaining corresponding water quality standards.

An equation for computation the assimilation capacity was described by Izrael, Tsiban, (1983) and Tsiban, Izrael (1988/1989):

$$A_m = Q_m \frac{C_{\text{threshold}}}{C_{\max}} \cdot v, \tag{5}$$

where Q_m is water volume in the considered area; $C_{threshold}$ and C_{max} – threshold and maximal concentrations of pollutant, respectively; v – averaged rate of the pollutant removal from the water column. One can see that such approach extends considerably the notion of environmental capacity, as it involves the rate of self-purification of seawater, which is expressed as:

$$v = \frac{\Delta C}{\Delta t}, \tag{6}$$

where ΔC – decreasing of the pollutant concentration in water column over the period Δt.

First who drew attention on the phenomenon and importance of the self-purification of aquatic environments was V.I. Vernadsky, proposing the theory of the unity of the processes of reproduction of the living matter and the conditioning of its habitat (Vernadsky, 1965). One of the further developments of this concept was the establishing of marine radiochemoecology, which deals with the study of dynamic aspects related to the input of pollutants into the seas and oceans, their migration between the various compartments, including kinetics of accumulation and transformation by biota, and deposition in the seabed sediments (Polikarpov, Egorov, 1981, 1986). Main result of these studies was the forming of concept of ecological capacity, which implies assessment of balance between the influxes and outflows of radioactive and chemical pollutants in the marine environments governed by the processes of self-purification (Egorov, 2001).

The following equations were developed to describe main biotic and abiotic factors controlling the self-purification of seawater and of other compartments marine environments against nuclear and non-nuclear pollutants:

period of turnover of radioactive or chemical substances in the living organisms:

$$T_B = \frac{1}{r + P + 1/T_{0.5}} \tag{7}$$

period of biotic turnover of radioactive or chemical substances in seawater:

$$T_W = \frac{1}{mK_B r} = \frac{K_m + C_W}{mV_{\max}} \tag{8}$$

transfer of radioactive or chemical substances through the food chains:

$$K_B = \frac{RqK_f}{Rq_f + r} \tag{9}$$

rate of biotic transformation of the pollutants:

$$V_{BT} = d_m C_W K_B m r = \frac{d_m m V_{\max} C_W}{K_m + C_W} \tag{10}$$

rate of self-purification of seawater due to the processes of biosedimentation:

$$V_{BS} = d_S P C_W K_B = \frac{d_S C_W P V_{\max}}{r(K_m + C_W)} \tag{11}$$

the limit of self-purification of seawater due to the processes of biosedimentation:

$$V_{BS}^{\max} = \frac{d_S P V_{\max}}{r}, \tag{12}$$

where m – specific biomass in seawater; K_B and K_f – concentration factors of the pollutants in a particular and previous links of the food chain, respectively; d_m – index of biotic transformation of toxic substances; d_S – fraction of primary production (P), which is removed from the water column due to the processes of sedimentation; r – specific rate of the pollutant exchange between the living organisms and seawater; $T_{0.5}$ – half-life of radionuclide decay or degradation of chemical pollutant; C_W – concentration of the pollutants in seawater; K_m – Michaelis-Menten constant; V_{max} – maximal rate of uptake of pollutant by the living organisms; R – specific diet of marine consumers; q – coefficient of the total uptake of the pollutant from food; q_f – coefficient the pollutants uptake from food for growth.

The flux of removal of the pollutants from water column is expressed as the rate of their deposition in the seabed sediments that is one of the main factors of immobilization of the toxic substances in marine environments (in case if no chemical analogous):

$$F_S = V_S C_S = V_S K_S C_W = \frac{V_S C_{\max} C_W}{1/k + C_W}, \tag{13}$$

where V_S – sediment accumulation rate; C_S – concentration of pollutant in sediments; C_{max} – maximal concentration of pollutant in sediments; K_S – concentration factor of the pollutant in the sediments; k – Langmuir adsorption constant.

Then, the thickness of the upper sediment layer, which accumulates the pollutant may be calculated as:

$$h = 5T_{0.5}bV_S, \tag{14}$$

where b – coefficient of the sediments compaction within this layer. And the upper limit of capability of sediments to accumulate pollutants is:

$$Q = C_{\max}h \tag{15}$$

Thus, equations (7-15) include parameters reflecting the metabolic characteristics of biota, ability of the living and non-living matter to concentrate the pollutants, productivity and sedimentation processes in ecosystem, and the capacity of seabed sediments to accumulate the radioactive and non-radioactive toxic substances, which are key components of the overall process of self-purification of the marine environments. The studies described in (Polikarpov, Egorov, 1986) have shown that if concentration of pollutant in seawater is much lower than metabolic and adsorptive constants ($C_W << K_m$ and $C_W << 1/k$) then its change depends linearly on the fluxes of contamination to the waters or seafloor sediments (Figure 1). Under the conditions of further increase of contamination inflow to the values when $C_W >> K_m$ and $C_W >> 1/k$ the flux of self-purification reaches the upper limit. Under these conditions, a steady state system: 'contamination–purification' discontinues. Both these situations may be described in terms of the contemporary theories of mineral and radioisotope exchange, as well as the sorption interactions, showing that metabolic and adsorptive reactions of the living and non-living matter may change under the increasing of concentration of the pollutant in seawater, from the first to zero order (Polikarpov, Egorov, 1986; Egorov, 2001).

The first order corresponds to the capability of water for purification, which is proportional to the concentration of the pollutants, due to the biosedimentation or biotransformation processes. At zero order of metabolic reactions, the increasing of concentration of the pollutant leads to the rising of the duration of its turnover but, in contrast, to the decreasing in ability for self-purification. If the input of toxic substance exceeds the upper limit, then the self-purification of seawater becomes unstable and no longer able to provide the conditioning of marine environments against pollution. This means that an important factor of maintaining the capacity of marine environments for self-purification is the adjustment of inflow of pollutants to the biogeochemical criteria reflecting accumulation and adsorptive functions of the living and non-living matters, the intensity of mass exchange between the neighboring waters, as well as the productivity and sedimentation processes in marine ecosystems.

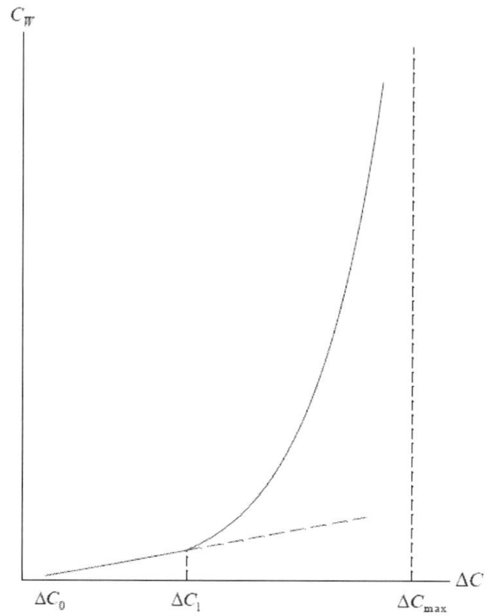

Figure 1. Concept of ecological capacity (after *Polikarpov, Egorov, 1981, 1986*): dynamics of concentration of pollutant in the upper seawater layer (C_W) depending on its influx (ΔC). ΔC_0 corresponds to the loss of the steady state; ΔC_1 – the loss of sustainable (quasi linear) growth; ΔC_{max} – the limit of the capacity for self-purification.

Half-Lives of Pollutants and the Self-Purification of Seawater

An integrated characteristic of the self-purification of seawater is the value of environmental half-life of radioactive and non-radioactive pollutants. Such the values combine the effective and ecological half-lives. The first value denotes the halving of concentration of nuclear and non-nuclear pollutant in a particular biotic or abiotic compartment of marine ecosystem by interaction of radioactive decay (or degradation of the chemical toxicant) and the processes of self-purification of the water column, while the ecological half-life implies the decrease in the pollutant concentration only due to the self-purification (e.g., Florou et al., 2002; Robison et al., 2003; IAEA, 2005; Nielsen et al., 2010; Kasamatsu, Inatomi, 2012).

One of the advanced opportunities for direct estimation of a 'net rate' of the self-purification and to differentiate between the effective and ecological half-lives is provided by data on time-series measurement of concentration of the man-made radionuclides in surface water of marine basins, which were subjected to radioactive contamination as a result of global fallout, particularly in the early 1960's, and/or of the local nuclear accident such as the Chernobyl disaster in April 1986. These fallouts were short-term, acting as a pulse input, and the only factor of radionuclide degradation is the radioactive decay, which has constant rate and independent of any environmental factors. Also, an additional feature of such the type of contamination is the possibility to identify quantitatively the main sources of the man-made radionuclides using their activity or atomic ratios, which are well documented for both releases: from the nuclear weapons testing and the accidental reactors (e.g., Aarkrog, 1994). All of these provide exceptional conditions for reliably interpreted tracking of the spatial and temporal trends of the seawater pollution, not only radioactive, but also chemical (e.g., IAEA, 2004; Osvath et al., 2007).

One of such estimations was carried out during the long-term radioecological studies and monitoring of concentration of the Chernobyl-derived radionuclides, particularly well-soluble ^{137}Cs and ^{90}Sr, and sorption-reactive plutonium isotopes, in the Black Sea, which owing to its geographical location and limited water exchange with the rest of the world's oceans has been one of the marine basins most contaminated with artificial radioactivity, ranking second after the Irish Sea and third after the Baltic Sea with respect to ^{90}Sr and ^{137}Cs concentration in seawater, respectively (e.g., IAEA, 2004; Egorov et al., 2010; Gulin et al, 2013). The data obtained have showed that post-Chernobyl evolution of man-made radioactivity in the Black Sea was governed, except radioactive decay, by the large-scale oceanographic, biogeochemical and ecological processes, which control the self-cleaning of the Black Sea environments against the radioactive contamination: first of all the vertical water mixing, which dilutes the surface radionuclide concentrations with the less contaminated deep-sea waters; the loss through the Bosporus Strait connecting the Black Sea with the Mediterranean; and the removal of particle-bound radionuclides on sinking fossils of plankton and suspended material including those come from the land due to erosion of soils within the watersheds. Because of the effect of these factors, as well as due to the radioactive decay, the initial inventory of the Chernobyl-derived radionuclides has decreased abruptly, reaching presently almost the pre-accident level, except estuarine and some near-shore zones, particularly in front of mouths of the Danube and Dnieper Rivers, which account for 75% of the total river runoff to the Black Sea. In turn, the study of the post-Chernobyl dynamics of radionuclide concentration in different compartments of the Black Sea ecosystem gave an advanced opportunity to evaluate a number of hydrological, geochemical, and ecological processes and to trace their long-term changes, achieving an integrated assessment of potential capability of the Black Sea for self-purification against both nuclear and non-nuclear pollutants (e.g., Polikarpov et al., 2008; Egorov et al., 2010; Gulin et al., 2012).

The summarized data on the temporal evolution of total inventory of the Chernobyl-derived ^{137}Cs in the upper Black Sea waters is given in Figure 2. It shows that initial concentration of ^{137}Cs in the surface layer (~2700 TBq) has decreased abruptly during one year period after the Chernobyl fallout due to a fast mixing of the water masses within the upper mixed layer, and then more gradually to around 500–600 TBq in 1998–2000 and 350 ± 60 TBq in 2001–2004 (Egorov et al., 2010; Gulin et al., 2012). Unlike ^{137}Cs, the decrease in ^{90}Sr inventory was not monotonic, showing considerable increase in 2001 (e.g. Egorov et al., 2010; Gulin et al., 2012, 2013). This may be explained by a larger ^{90}Sr inflow with the river waters due to its higher mobility in the freshwater environments and because of a stronger runoff occurring in the Dnieper and Danube rivers in 1998–1999. In contrast, the ^{137}Cs inflow from the rivers was negligible (<1.5% of its total post-Chernobyl inventory in the Black Sea) because this radionuclide is sorption reactive in freshwater and therefore stronger retained in the watersheds (e.g., Martin et al., 1994; Thomas, 1997; Gulin et al., 2002).

The exponential function approximating the time-series measurements of ^{137}Cs concentration in the upper 50-m Black Sea water resulted in the effective half-lives for decrease of ^{137}Cs inventory between 5.3 years in 1987–2000 (Egorov et al., 2010), 7.0 years in 1987–2004 (Egorov et al., 2010) and 8.6 years over the period of 1987-2011 (Gulin et al., 2013) or 6.6 years on the average. The gradual increase of these half-lives with time may be caused presumably by growing influence of remobilization of ^{137}Cs from the seabed sediments and due to a slow migration from watersheds of the adsorbed ^{137}Cs with the eroded soil particles (Gulin et al., 2002; Gulin et al., 2003; Gulin et al., 2013). This assumption is supported additionally by the larger values of the half-life for decrease in the surface Black Sea water of concentration of ^{90}Sr, well-soluble in both riverine and seawater, which was varying from 8.3 years over 1986–2000 period to 12.6 years in 1986–2005 due to the considerable secondary input of this Chernobyl radionuclide from rivers (Egorov et al., 2010; Gulin et al., 2013).

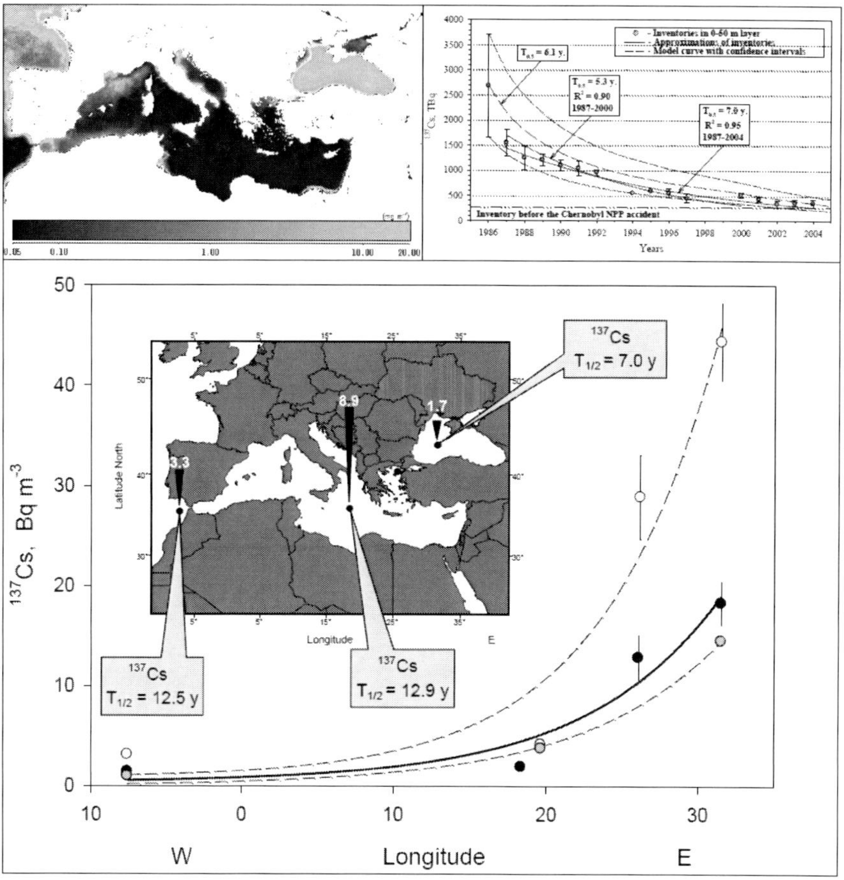

Figure 2. Longitudinal distribution of ^{137}Cs and $^{239+240}$Pu (black inverted triangular bars in the map) concentration in surface waters of the western Black Sea, Aegean Sea (near Dardanelles), Ionian Sea and in the Atlantic area adjacent to the Gibraltar Strait measured in 2002–2003 (●) in comparison with the post- (○) and pre-Chernobyl (gray spots) data obtained in 1988–89 and the early 1980's, respectively. Error bars are shown if greater than symbol size. Numbers above the black triangular bars related to the values of $^{239+240}$Pu concentration in mBq·m^{-3}. The left upper panel shows satellite map of annual mean surface chlorophyll concentration for 2002 (the grey-scale coding shows pigment concentration in mg·m^{-3}). The right upper panel represents temporal evolution of ^{137}Cs inventory in 0–50 m layer of the Black Sea after the Chernobyl NPP accident.

The given value of effective half-life for ^{137}Cs in the Black Sea surface water is considerably shorter than those found by us earlier in the central part of Mediterranean (mid-Ionian Sea) and in the NE Atlantic area adjacent to the Gibraltar Strait (Figure 2), where half-lives for ^{137}Cs in the surface water were 12.9±1.0 and 12.5±2.1 years, respectively (Gulin et al., 2012; 2013). The lesser half-life, which was found in the Black Sea, is most likely caused by a faster dilution of ^{137}Cs in the surface seawater with riverine waters, as well as by a loss of ^{137}Cs via the Bosporus Strait connecting the Black and Marmara Seas.

Also, the higher self-purification rate of the Black Sea waters, relative to Mediterranean and NE Atlantic, has been revealed for particle-reactive $^{239+240}$Pu, whose surface water concentration measured in 2002 was considerably lower in the Black Sea than in Ionian Sea (Figure 2) despite the latter is being located much farther away from the accident site. These data, in comparison with results of the previous measurements of plutonium concentration in the Black Sea waters (e.g., Egorov et al., 2010), suggest the effective half-life of $^{239+240}$Pu of ~6.5 years on the average.

In this case, the rapid decrease of $^{239+240}$Pu surface water concentration is most likely caused by a higher rate of plutonium scavenging on sinking particles in the mesotrophic Black Sea basin, in contrast to oligotrophyc Mediterranean waters (Gulin et al., 2012; 2013).

A wider-scale measurements of the effective half-lives were carried out using the data on latitudinal distribution of ^{137}Cs in surface waters along a transect crossing the NE Atlantic – western Antarctic waters during the 7th Ukrainian Antarctic Expedition in 2001-2002 (Gulin, Stokozov, 2005). Comparison of these data with results of previous same-route expeditions SWEDARP (Swedish Antarctic Research Expedition) in 1988/1989 and of French R/V "Jeanne d'Arc" in 1992/1993 (Bourlat et al., 1996; Holm et al., 1996) has shown practically parallel changes of ^{137}Cs surface concentrations between 40° N and 20° S, pointing to decrease of ^{137}Cs radioactivity at these latitudes with the effective half-life of 10-15 years (12.5±2.1 years, Figure 3).

Since the obtained half-lives are much shorter than the ^{137}Cs radioactive half-life (30.17 years), this suggests that the decrease of ^{137}Cs concentration in surface waters within this latitude band was controlled, besides the radioactive decay, by vertical mixing of the upper water masses with water from the deeper layers (Gulin, Stokozov, 2005).

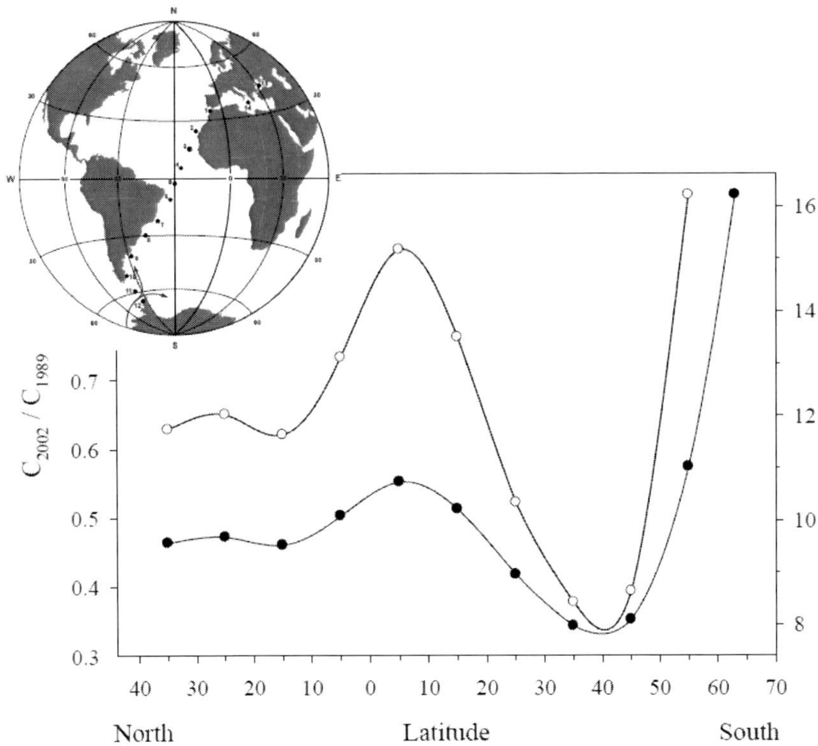

Figure 3. Relative changes of ^{137}Cs concentrations in Atlantic/Antarctic surface waters (closed circles) measured during the SWEDARP cruise in 1988/89 (C_{1989}) and the 7th Ukrainian Antarctic Expedition in 2001/2002 (C_{2002}). Open circles show the effective half-life of ^{137}Cs at the respective latitude bands. The upper inset represents the map of the sampling sites along the route of 7th Ukrainian Antarctic expedition. Arrows show a part of the circumpolar current flowing around Antarctica and the northward Falkland current.

South of 20° S, the ^{137}Cs concentrations in surface water have decreased from 1988/1989 more rapidly, most likely because of influence of the Falkland current bringing the less contaminated Antarctic waters to the given area, and also due to a higher vertical mixing of surface water at the storm-latitudes within 40° S where the apparent half-life of ^{137}Cs is shortest (~8 years, Figure 3).

The differentiation between the effective and ecological half-lives may be achieved using the simple relationship:

$$T_{Eff} = \frac{T_{0.5} \cdot T_{Ecol}}{T_{0.5} + T_{Ecol}}, \tag{16}$$

where T_{Eff} – effective half-life; T_{Ecol} – ecological half-life; $T_{0.5}$ – half-life due to the radioactive decay.

The change of radionuclide concentration with time in the upper water column may be described by the following differential equation:

$$\frac{dC_W}{dt} = pC_W - \lambda C_W - s_{Ecol} C_W + rC_W = C_W (p - \lambda - s_{Ecol} + r), \tag{17}$$

where constant p defines the specific rate of the input of pollutant; $\lambda = \ln 2/T_{0.5}$ – constant of radioactive decay; $s_{Ecol} = \ln 2/T_{Ecol}$ – specific rate of the decrease in radionuclide concentration with time due to the processes of self-purification; r – specific rate of the radionuclide remobilization from seabed sediments and resulting from degradation of suspended matter or decomposition of dead organisms.

One may readily find the solution of equation (17) to be:

$$C_W(t) = C_{W_O} e^{k \cdot t}, \tag{18}$$

where $k = p - \lambda - s_{Ecol} + r$; C_{Wo} – initial concentration of radionuclide in seawater.

Since the rates of self-purification of seawater against the suspended and dissolved forms of the pollutants may be differ, the dynamics of its total concentration is to be expressed as:

$$\frac{dC_W}{dt} = \frac{d(C_p + C_d)}{dt} = C_p k_p + C_d k_d = aC_W k_p + (1-a) C_W k_d \tag{19}$$

and

$$C_W(t) = C_{p_0} e^{k_p \cdot t} + C_{d_0} e^{k_d \cdot t} = aC_{W_0} e^{k_p \cdot t} + (1-a) C_{W_0} e^{k_d \cdot t}, \tag{20}$$

where C_p and C_d – concentrations in seawater of particulate and dissolved fractions of the pollutant; k_p and k_d – constants of the rate of their changes with time *t*; *a* – percentage (fraction) of the pollutant in particulate form. An example of computation, using the equations (16) and (18) and the most recent data on ^{137}Cs concentration in the upper Black Sea waters (*Gulin et al., 2013*), is given in Figure 4.

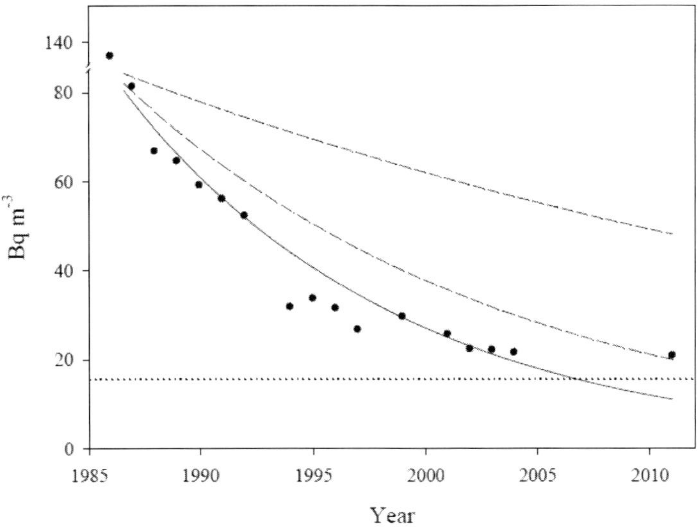

Figure 4. Decreasing of ^{137}Cs concentration in the upper Black Sea water after the Chernobyl accident (●). **Solid line computed for** effective half-life of ^{137}Cs. The upper dashed line represents the radioactive decay of ^{137}Cs. The lower dashed line calculated for ecological half-life of ^{137}Cs. The doted line indicates the pre-accident level of ^{137}Cs concentration.

It shows significant contribution of self-purification to the decrease of ^{137}Cs concentration in the upper Black Sea waters (average T_{Eff} = 6.6 years, see above), suggesting that in case of the absent of self-purification, the 90% level of decrease would be reached, at the account only of radioactive decay, in around 100 years after the fallout, whereas the actual duration of this period is more than 3 times shorter (Figure 4). Assuming no significant contribution of remobilization of ^{137}Cs from the seabed sediments, particularly during the first years after the Chernobyl-derived fallout (when effective half-life for this well-soluble radionuclide in the upper Black Sea water was estimated to be 5.3 years – see above), gives the value of ecological half-life of 6.4 years and the respective constant of a net specific rate of self-purification of 0.11 yr^{-1}. This value is independent of the contribution of radioactive decay and can be used, therefore, with regards to a wide range of non-nuclear pollutants too. If to set that pollutant concentration can not exceed its maximum allowable concentration (MAC):

$$C_W(t) \le MAC, \qquad (21)$$

then specific rate of the *maximum allowable input* of the pollutant (MAI) over one year period ($t = 1$) may be determined as:

$$p = \ln\left(\frac{MAC}{C_{W_o}}\right) + \lambda + s_{Ecol} - r \qquad (22)$$

If to apply, as the measure of self-purification rate, the above-mentioned values of the effective half-lives of the decrease in concentration of well-soluble and high particle-reactive fallout radionuclides (6.5-6.6 years on the average in both ceases, see above), then we can express the integrated value of specific rate of the self-purification by combining its with specific rates of decay and remobilization:

$$s_{Eff} = s_{Ecol} + \lambda - r \qquad (23)$$

This allows simplification of the equation (22) as follows:

$$p = \ln\left(\frac{MAC}{C_{W_o}}\right) + s_{Eff} \qquad (24)$$

The given value shows that specific rate of the input of pollutant depend on the ratio between its maximum allowable concentration and initial level of the pollution in water, and that the upper limit of the rate of the pollutant input may increase with the respective increasing of the self-purification rate. Thus, this parameter combines characteristics of toxicity of pollutant and an integrated rate of the self-purification of seawater. It, therefore, could be applied for purposes of environmental regulation of the maximum allowable anthropogenic pressure with regards to the nuclear and non-nuclear pollutants, applying the value of specific rate of the *maximum allowable annual input* of the pollutant, at which its concentration will not exceed the MAC level over one year period. But after that period, according to the equations (18) or (20), concentration of pollutant will continue to increase progressively. In order to avoid this, an upper limit of the pollutant concentration, equals the MAC, should be included in equation (17), e.g. by transforming it in the logistic sigmoid (Verhulst) function:

$$\frac{dC_W}{dt} = kC_W \cdot \left(1 - \frac{C_W}{MAC}\right), \tag{25}$$

where

$$k = p - s_{Eff}, \tag{26}$$

assuming that specific rates of self-purification of seawater against the suspended and dissolved pollutant are comparable (i.e. $k_p \approx k_d$ as it was in the above-mentioned case with the decreasing of concentration of the well-soluble ^{137}Cs and particle-reactive $^{239+240}Pu$).

Equation (25) implies that if concentration of pollutant is much lower than MAC, then its dynamics may be described as the exponential growth similarly with equation (17), and, therefore, the value p, given in equation (24), can be applied also in equations (25-26). But the closer this concentration is approaching the upper allowable level, the slower it is rising and stops completely ($dC/dt = 0$) when C_W has reached the MAC for infinite time (or come closer to reaching in finite time).

It is important to stress that this model corresponds to a case when self-purification of the water body is not sufficient itself to maintain the pollutant concentration at the level not exceeding the MAC, and therefore an additional restriction of the discharge of the polluting substance is required as a mean to avoid the excess contamination.

The solution of the logistic model (25) is:

$$C_W(t) = \frac{C_{W_O} \cdot MAC}{C_{W_O} + (MAC - C_{W_O}) \cdot e^{-k \cdot t}} \quad or \quad C_W(t) = \frac{C_{W_O} \cdot MAC \cdot e^{k \cdot t}}{MAC - C_{W_O} + C_{W_O} \cdot e^{k \cdot t}} \tag{27}$$

As can be seen from an example of the logistic curve plotted in Figure 5, the beginning of this function shows exponential rise, which is slowing to the linear growth and then approaches the upper limit with an exponentially decaying curve.

The first derivative of the function (27) is:

$$C_W(t)' = \frac{k \cdot MAC \cdot C_{W_O} \cdot e^{k \cdot t} \cdot (MAC - C_{W_O})}{(MAC - C_{W_O} + C_{W_O} \cdot e^{k \cdot t})^2} \tag{28}$$

This derivative is positive if $C_{Wo} < MAC$ and negative if $C_{Wo} > MAC$. This means that in the first case the function rises gradually, reaching the value of the maximum allowable concentration, and decreases monotonically (at $t > 0$) down to the MAC-level in case when the initial concentration of the pollutant was higher than MAC (Figure 5).

The second derivative of the function (27) is:

$$C_W(t)'' = \frac{k^2 \cdot MAC \cdot C_{W_O} \cdot e^{k \cdot t} \cdot (MAC - C_{W_O})}{(MAC - C_{W_O} + C_{W_O} \cdot e^{k \cdot t})^3} \cdot (MAC - C_{W_O} - C_{W_O} \cdot e^{k \cdot t}) \tag{29}$$

This equation shows that if $C_{Wo} > MAC$, the value $C_W(t)'' \neq 0$ and the function does not have any inflection points, but if $C_{Wo} < MAC$, then C(t)" = 0 when $MAC - C_{Wo} - C_{Wo} \cdot e^{k \cdot t} = 0$. This point of inflection corresponds to the time period when the rate of maximum allowable input of the pollutant has reached the peak (see the bell-shaped curve dC_W/dt in Figure 5):

$$t_{0.50} = \frac{1}{k} \cdot \ln\left(\frac{MAC - C_{W_O}}{C_{W_O}}\right) \tag{30}$$

Accordingly, the value of the function (27) at this time point is half of the MAC:

$$C_W(t_{0.50}) = \frac{MAC \cdot C_{W_O} \cdot e^{k \cdot t_{0.50}}}{MAC - C_{W_O} + C_{W_O} e^{k \cdot t_{0.50}}} = \frac{MAC}{2}, \tag{31}$$

This means that the rate of changes of the pollutant concentration (dC_W/dt) will increase up to the time = $t_{0.50}$ when the concentration is ½ of MAC and then will be decreasing down to $dC_W/dt = 0$ at time point when the concentration reaches the MAC-level (Figure 5).

According to equation (27), the time period, over of which the value $C_W(t)$ will reach asymptotically the 99% level of the MAC (i.e. when C_W = 0.99MAC), may be assessed as:

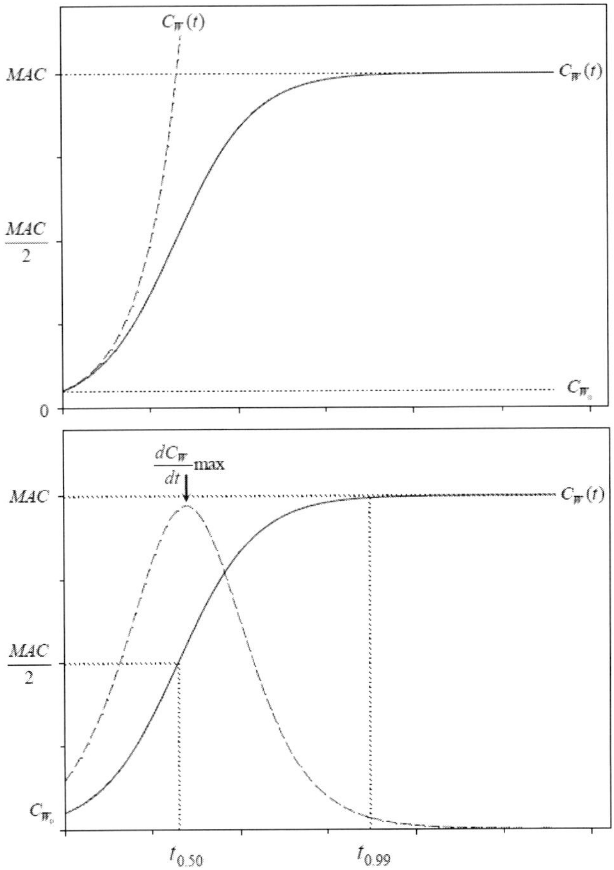

Figure 5. Exponential (top panel, dashed line) and logistic (both panels, solid line) growth of the pollutant concentration in seawater and the dynamics of its change – dC_W/dt (bottom panel, dashed bell-shaped line).

$$t_{0.99} = \frac{1}{k} \cdot \ln\left(\frac{99 \cdot (MAC - C_{W_O})}{C_{W_O}} \right) \tag{32}$$

This equation allows solution of inverse problem of justifying the specific rate of the pollutant input, so that its concentration would reach the MAC level over the predetermined period of time, e.g. in one year ($t_{0.99} = 1$):

$$k = \ln\left(\frac{99 \cdot (MAC - C_{W_O})}{C_{W_O}}\right), \quad p = k + s_{Eff} \tag{33}$$

The given *p*-value represents an adjusted estimate of specific rate of *maximum allowable annual input* of the pollutant to be used as a practicable measure for environmental control of contamination, that prevents the exceeding over the maximum admissible limits.

Finally, the basin-wide environmental capacity of the respective layer of seawater may be computed using the integration:

$$A_{\max} = \frac{1}{t_{0.99}} \cdot V \cdot \int_0^{t_{0.99}} \left(C_W(t) - C_{W_O}\right) dt, \tag{34}$$

where V – volume of water in the considered layer.

An example of such the calculations can be illustrated regarding the pollution of the Black Sea surface water with mercury using our most recent data obtained during the 70th cruise of RV "Professor Vodyanitskiy" in 2011. According to these data, the highest concentration of mercury in the upper seawater layer was of 83 $ng \cdot L^{-1}$, while the basin-wide average value (to be used as C_{W0} value) was much lower, 18 $ng \cdot L^{-1}$. The maximum allowable concentration (MAC) of mercury, which is set for natural surface waters, is varied in different Black Sea countries around of 100 $ng \cdot L^{-1}$ (e.g., UNEP, 2002). Applying these values, as well as the above-mentioned value of ecological half-life s_{Eff} = 6.4 years derived from radionuclide data, gives from equation (24) the value of specific rate of the maximum allowable annual input of mercury (p) of 4.5 yr^{-1}. According to the equations (30) and (32), the time period when the mercury concentration will reach the half of the MAC at such the rate of the input ($t_{0.50}$) is 0.34 years, whereas the time needs to approach the 99% level of MAC ($t_{0.99}$) ≈ 1.4 years. As it can be seen at the bell-shaped plot of the function dC_W/dt in Figure 5, and taking into account the volume of the upper 50-m layer of the Black Sea water (in which the self-purification rate was determined using the radionuclide data, see Figure 2) ≈ $2.16 \cdot 10^{13}$ m^3 (e.g., Bogdanova, 1959), the rate of the mercury input may gradually increase up to the maximal level of ~$1.4 \cdot 10^3$ $t \cdot yr^{-1}$ corresponding to the time $t_{0.50}$ when the mercury concentration will reach the half of the MAC (Figure 5). After this period, the rate of the annual input of mercury to the surface Black Sea waters must be steadily reduced down to the full stop when its concentration will come closer to reaching the MAC level in $t_{0.99}$ years (Figure 5).

It should be noted, however, that above-mentioned maximal rate of the mercury input ($1.4\cdot10^3$ $t\cdot yr^{-1}$) is the highest admissible value, which will leads to the rising of the mercury concentration to the level of its maximum allowable concentration (100 $ng\cdot L^{-1}$) within the whole 50-m thickness of the upper Black Sea waters. But, if to aim the keeping a steady state of the contamination (when $dC_W/dt=0$), then the maximum input, at which its concentration will maintain at the current level $C_W(t)=C_{W0}=const=18$ $ng\cdot L^{-1}$, may be estimated as a value inverse of the rate of self-purification using the formula derived from equations (17) and (19), assuming the similarity of specific rates of self-purification of the dissolved and suspended fractions of mercury (i.e. $k_d \approx k_p$), as it occurs with well-soluble and particle-reactive Chernobyl radionuclides (see above):

$$p\cdot C_{W_0} = s_{Eff}\cdot C_{W_0} = \frac{\ln 2}{T_{Eff}}\cdot C_{W_0} = \frac{\ln 2}{6.4\,yr}\cdot 18\,ng\cdot L^{-1} = 1.98\,ng\cdot L^{-1}\cdot yr^{-1} \qquad (35)$$

Applying the above-mentioned volume of the upper 50-m layer of the Black Sea water yields the basin-wide 'steady-state-providing' assimilation capacity of this water mass to be nearly 40 $t\cdot yr^{-1}$. This value fits well to the range of previous indirect assessments of such the capacity of the Black Sea waters regarding its contamination with mercury, which varied between 20 and 600 $t\cdot yr^{-1}$ (Polikarpov, Egorov, 1986, and references therein). It should be emphasized, however, that the given value of relatively 'safe' mercury input has already became comparable with the annual inflow of this high-toxic heavy metal from the Danube River, which is estimated as 60 $t\cdot yr^{-1}$ (e.g., EU Parliamentary Assembly, Committee on the Environment, Doc. 11632, 2008), pointing the necessity of an immediate reduction of amount of the mercury discharge into the Black Sea basin.

Thus, the data on temporal evolution of the man-made radioactivity in seawater provides opportunity to evaluate directly the levels of maximum allowable discharge of pollutants, both nuclear and non-nuclear, into marine environments and to assess their capacity regarding the contamination to be used as quantitative measure for environmental regulation of human impacts allowing to maintain it at a steady state or at the level not exciding the MAC.

There is, however, an issue concerning the definition of the MAC levels, which may vary from one country to another due to different environmental standards and policy. This is particularly important for comparative assessment of the influence of nuclear and non-nuclear contaminants having different nature of biological effects and different dimension of the measuring units. An approach for such the comparison of equivalent effects of the diverse stressors in aquatic ecosystems was originated by Polikarpov (1998), who developed a conceptual model of responses of the living organisms, their populations, communities and the whole ecosystems to all possible dose rates of ionizing radiation in the environment with the following zones of radiation effects: radiation well-being; physiological masking; ecological masking; and damage to ecosystems. This model was then extended into the radiochemoecological equi-dosimetric model, which covers not only ionizing exposure but also the effects of chemical pollutants and other types of anthropogenic impacts with purpose of ranking of their ecotoxicity on the basis of dosimetric equivalents, e.g. Gray-equivalent expressed as dose rates $Gy \cdot yr^{-1}$ (Polikarpov, 1998, 1999, 2001; Fuma et al., 1998; Miyamoto et al., 1998; Takeda et al., 1998). As an example of the applicability of this approach, it was shown that anthropogenic pressure upon the unique Black Sea field of algae Phylophora, 97% of whose area were lost during the last 30 years due to eutrophication, pollution and the increase in water turbidity, and the radiation effects found in the well-known “Red forest” located in the Chernobyl restricted zone, are almost equal in units of the Gy-equivalents.

But this approach would require every time to carry out the parallel ecotoxicological and radioecological dose-effect experiments that is not always accessible. In order to overcome this difficulty, a dimensionless value expressed as a fraction (or percentage) of maximum resistance capability (MRC) of marine organisms to radiation or toxic exposures has been recently proposed, and still is being developed, in (Gulin, 2013) based upon experiments on the effects of incorporated radionuclides with different biological significance (^{14}C, ^{106}Ru, ^{144}Ce) and mercury, under the small, medium and lethal doses, on the division rate and dimensional characteristics of marine unicellular algae. These experiments showed a similar shape of the dose-response relationships, which included the phase of initial uncertainty under a lowest radiation exposure or mercury concentration, and then increasing of the cells division rate as compared to the non-exposed algae up to the threshold dose/concentration at which the cells began to die. This threshold level is proposed to define as the maximum resistance capability (MRC) of marine biota, that equals to the uppermost allowable level of toxic impact, suggesting to apply a dimensionless fraction values of the MRC as a comparable measure of the impact of both radioactive and chemical substances.

Finally, it is obvious that the rate of self-purification of seawater may be not always constant. According to contemporary theories, the functioning of ecosystems may be described as a combination of direct and feedbacks using the principle of Le Châtelier and Braun which implies that biogeocoenoses can be in the state of resistance or compensating homeostasis (e.g., Odum, 1983). In marine ecosystems, the resisting homeostasis is observed at relatively low level of anthropogenic pressure when the compensating capability is provided by their structure (Egorov, 2012).

Under this regime of homeostasis the main purpose of environmental regulation is to define the maximum allowable influxes of pollutants, which can be balanced by processes of self-purification of seawater under conditions when the steady-state concentration of toxic substances does not exceed the MAC levels (see above). But anthropogenic impact can influence, in turn, on the rate of the self-purification of marine ecosystems through the changes of their biodiversity and productivity. This leads to the transition to a compensatory homeostasis implying the respective changes in capability of the ecosystems for the self-decontamination.

For instance, the increase of anthropogenic impact in the Black Sea ecosystems after 1970's, as well as the climatic changes and the invasion organisms from other regions of the World ocean (e.g., Black Sea Transboundary Diagnostic Analysis, 1997), resulted in increase of primary production, in reducing the size of phytoplankton cells and of contribution of macro algae to the total productivity, in increase of bacterial production and the decrease of abundance and sizes of zooplankton and fish (e.g., Zaitsev, Mamaev, 1997). These changes have led to the increase in concentration of biogenic particles in the Black Sea water column, including plankton fossils and fecal pellets, and to the following faster scavenging of particle-bound radionuclides and chemical pollutants.

Hence, in can be concluded that in order to solve the problem of scientifically grounded environmental regulation of anthropogenic pressure on marine ecosystems it is necessary to assess their ability for compensative homeostasis under the condition of the changeable rate of self-purification of seawater.

Conclusion

The main problems and needs that have to be addressed in the context of environmental regulation of pollution of marine environments are: (1) incompleteness, poor comparability or absence of the environmental monitoring data for radioactive and non-radioactive contaminants; (2) a lot of gaps in the existing datasets regarding the most of critical pollutants, particularly of those which were released to the global and regional environments long before the development of reliable methods of their detection in marine ecosystems; (3) uncertainty in the pollution trends that may lead to significant loss of quality of assessments of the current environmental state of the marine ecosystems, of differentiation between the human- and climate-induced processes driving the changes in marine environments, and of regional environment prediction models; (4) lack of quantitative assessments of capability of the marine ecosystems for self-decontamination and of the respective maximum allowable pollution loads; (5) weakness of criteria for comparative assessment of environmental state of marine ecosystems subjected to different-type anthropogenic impacts.

One of the promising ways to solve these problems is the radiochemoecological evaluation of capacity of marine environments for self-purification against radioactive and non-radioactive pollutants. Data on temporal evolution of the man-made radioactivity in seawater provides opportunity to evaluate directly the levels of maximum allowable discharge into marine environments of nuclear and non-nuclear pollutants, to differentiate between their effective and ecological half-lives, and to assess capacity of aquatic ecosystems regarding the assimilation of contamination to be used as quantitative measure for environmental regulation of human impacts allowing to maintain it at a steady state or at the level not exciding the maximum allowable concentration.

The developed criterion of the maximum allowable annual input of the toxic contaminants may be a useful tool for environmental regulation of pollution of seawater with the nuclear and non-nuclear substances at a common methodological basis.

In order to solve the problem of scientifically justified environmental regulation of anthropogenic pressure on marine ecosystems it is necessary to assess their ability for compensative homeostasis under the condition of the changeable rate of self-purification of seawater.

REFERENCES

Aarkrog, A. (1994). Source terms and inventories of anthropogenic radionuclides. In: E. Holm (Ed.), *Radioecology. Lectures in Environmental Radioactivity* (pp. 21–38). Lund (Sweden), World Scientific Publishing.

Agre, A. L., Korogodin, V. I. (1960). On the distribution of radioactive contamination in the non-flow-through water body, *Medical Radiology, 1*, 67–73 (in Russian).

Black Sea Transboundary Diagnostic Analysis. (1997). BSEP/ UNDP, New York.

Bogdanova, A. K. (1959). Water exchange through the Bosporus Strait and its role in the mixing of the Black Sea waters, *Proceedings of the Sevastopol Biological Station, 12*, 401–420 (in Russian).

Bourlat, Y., Milliès-Lacroix, J.-C., Le Petit, G., Bourguignon, J. (1996). ^{90}Sr, ^{137}Cs and $^{239+240}Pu$ in Word Ocean water samples collected from 1992 to 1994. In: P. Guéguéniat, P. Germain, H. Métivier (Eds.), *Radionuclides in the Oceans: Inputs and Inventories* (pp. 75–93). IPSN, Editions de Physique.

Bronfman, A. M. (1992). Self-purification in the context of the problems of anthropogenic ecology of the sea, *GeoJournal,* 27(2), 141–148.

Egorov, V. N. (2001). Regulation of fluxes of anthropogenic pollution of the Black Sea regions using the biogeochemical criteria, *Ecology of Sea,* 57, 75–84 (in Russian).

Egorov, V. N. (2012). Biogeochemical mechanisms of realization of compensatory homeostasis in the Black Sea ecosystems, *Marine Ecological Journal,* 4, 5–17 (in Russian).

Egorov, V. N., Povinec, P. P., Polikarpov, G. G., Stokozov, N. A., Gulin, S. B., Kulebakina, L. G. & Osvath, I. (1999). ^{90}Sr and ^{137}Cs in the Black Sea after the Chernobyl NPP accident: inventories, balance and tracer applications. *Journal of Environmental Radioactivity,* 49 (3), 137–156.

Egorov, V. N., Gulin, S. B., Polikarpov, G. G., Osvath, I. (2010). Black Sea. In: D. A. Atwood (Ed.), *Radionuclides in the Environment* (pp. 430–452). John Wiley & Sons, Ltd, Chichester (UK).

EC. (1996). Technical Guidance Document in support of the Commission Directive 93/67/EEC on *Risk Assessment for New Notified Substances and Commission Regulation* (EC) No 1488/94 on Risk Assessment for Existing Substances. Parts 1-4. Office for Official Publications of the EC, Luxembourg.

Florou, H., Kritidis, P., Vosniakos, F., Polikarpov, G., Egorov, V., Delfanti, R., Papucci, C., Aladjadjiyan, A. (2002). Dispersion of ^{137}Cs in the eastern Mediterranean and the Black Sea: the time evolution in relation to the sources and pathways. *Journal of Environmental Protection and Ecology,* 3 (1), 30.

Fuma, S., Miyamoto, K., Takeda, H. (1998). Ecological effects of radiation and other environmental stress on aquatic microcosm. In: J. Inaba, Y. Nakamura (Eds.), *Comparative Evaluation of Environmental Toxicants – Health Effects of Environmental Toxicants Derived from Advanced Technologies* (pp. 131–144). Kodansha Scientific, Ltd, Tokyo.

GESAMP. (1982). The review of the health of the oceans. In: *Reports and Studies. GESAMP, 15*, UNEP, New York.

GESAMP (1990). The state of the marine environments. In: *Reports and Studies. GESAMP, 39*, UNEP, New York.

Gulin, S. B., Polikarpov, G. G., Egorov, V. N., Zherko, N. V., Stokozov, N. A. (1995). Chronological study of ^{137}Cs, PCBs and some pesticides fluxes into the western Black Sea deep sediments. In: *Radiation Protection-70. The radiological exposure of the population of the European Community to radioactivity in the Mediterranean Sea (MARINA-MED Project): Proc. of a seminar held in Rome at the European Nuclear Energy Agency headquarters (17-19 May 1994)*, Brussels, Luxembourg, 487–500.

Gulin, S. B., Aarkrog, A., Polikarpov, G. G., Nielsen, S. P. & Egorov, V. N. (1997). Chronological study of ^{137}Cs input to the Black Sea deep and shelf sediments, *Radioprotection, 32* (NC2), 257–262.

Gulin, S. B., Polikarpov, G. G., Egorov, V. N., Martin, J. M., Korotkov, A. A. & Stokozov, N. A. (2002). Radioactive contamination of the north–western Black Sea sediments, *Estuarine, Coastal and Shelf Science, 54* (3), 541–549.

Gulin, S. B., Egorov, V. N., Polikarpov, G. G., Stokozov, N. A., Mirzoyeva, N. Yu., Tereschenko, N. N., Osvath, I. (2012). General trends in radioactive contamination of the marine environment from the Black Sea to Antarctic Ocean. In: E. B. Burlakova, V. I. Naydich (Eds.), *The Lessons of Chernobyl: 25 Years Later* (pp. 281–299). Nova Science Publishers, New York.

Gulin, S. B., Polikarpov, G. G., Martin, J.-M. (2003). Geochronological reconstruction of ^{137}Cs transport from the Coruh river to the SE Black Sea: comparative assessment of radionuclide retention in the mountainous catchment area, *Continental Shelf Research, 23*, 1811–1819.

Gulin, S. B., Stokozov, N. A. (2005). ^{137}Cs concentrations in Atlantic and western Antarctic surface waters: results of the 7th Ukrainian Antarctic Expedition, 2002, *Journal of Environmental Radioactivity, 83, 1*, 1–7.

Gulin, S. B. (2013). Effect of ionizing radiation and mercury on the unicellular algae: radiation-ecotoxicological study, *Marine Ecological Journal, 2*, 1–13 (in Russian).

Gulin, S. B., Mirzoyeva, N. Yu., Egorov, V. N., Polikarpov, G. G., Sidorov, I. G., Proskurnin, V. Yu. (2013). Secondary radioactive contamination of the Black Sea after Chernobyl accident: recent levels, pathways and trends, *Journal of Environmental Radioactivity, 124*, 50–56.

Holm, E., Roos, P., Josefsson, D., Persson, B. (1996). Radioactivity from the North Pole to the Antarctic. In: P. Guéguéniat, P. Germain, H. Métivier (Eds.), *Radionuclides in the Oceans: Inputs and Inventories* (pp. 59–74). IPSN, Editions de Physique.

IAEA. (2004). *Regional Technical Co-operation Project RER/2/003 "Marine Environmental Assessment in the Black Sea Region": Project report*, Vienna: IAEA.

IAEA. (2005). *Worldwide marine radioactivity studies (WOMARS). Radionuclide levels in oceans and seas*. IAEA-TECDOC-1429. Vienna: IAEA.

Izrael, Yu. A., Tsiban, A. V. (1983). About assimilation capacity of the World ocean, *Proceedings of Academy of Sciences of USSR*, 272(3), 702–705 (in Russian).

Kasamatsu, F., Inatomi, N. (2012). Effective environmental half-lives of ^{90}Sr and ^{137}Cs in the coastal seawater of Japan, *Journal of Geophysical Research: Oceans, 103(C1)*, 1209–1217.

Korogodina, V. L., Korogodin, V. I., Kutlakhmedov, Yu. A. (1996). Radiocapacity: prognosis of pollution after nuclear accidents, *Proc. IX Int. Congress on Radiation Protection (IRPA9),* Vienna: IAEA.

Leandri, M. (2009). The shadow price of assimilative capacity in optimal flow pollution control, *Ecological Economics, 68*, 1020–1031.

Martin, J. M., Wollast, R., Loijens, M., Thomas, A., Mouchel, J. M., Nieuwenhuize, J. (1994). Origin and fate of artificial radionuclides in the Scheldt estuary, *Marine Chemistry, 46*, 189–202.

Miyamoto, K., Fuma, S., Takeda, H. (1998). Effect of acidification on the population of growth stage aquatic microcosm. In: J. Inaba, Y. Nakamura (Eds.), *Comparative Evaluation of Environmental Toxicants – Health Effects of Environmental Toxicants Derived from Advanced Technologies* (pp. 127–130) Kodansha Scientific, Ltd, Tokyo.

Muller, F. L. L., Gulin, S. B., Kalvøy, Å. (2001). Chemical speciation of copper and zinc in surface waters of the western Black Sea, *Marine Chemistry, 76*, 233–251

Nielsen, S.P., Lüning, M., Ilus, E., Herrmann, J., Outola, I., Ikäheimonen, T., et al., 2010. Baltic Sea. In: D. A. Atwood (Ed.), *Radionuclides in the Environment* (pp. 415–436). John Wiley & Sons, Ltd, Chichester (UK).

Odum, E. (1983). Basic Ecology. Saunders College, Philadelphia.

Osvath, I., Egorov, V., Gulin, S., Laptev, G., Voitsekhovitch, O. (2007). What can radiotracers tell us about the fate of Black Sea contaminants? *Rapp. du 38-e de la SIESM, 38*, 20.

Pearce, D. W., Turner, R. K. (1990). *Economics of natural resources and the environment*. Johns Hopkins University Press, Baltimore.

Polikarpov, G. G. (1966). *Radioecology of Aquatic Organisms*, Reinhold, New York.

Polikarpov, G. G. (1999). Effects of nuclear and non-nuclear pollutants. In: *Marine Pollution. – Proceedings of Symposium, Monaco, 5–9 October 1998* (pp. 38–43). IAEA-TECDOC-1094, IAEA, Vienna.

Polikarpov, G. G. (2001). The future of radioecology: in partnership with chemo-ecology and eco-ethics, *Journal of Environmental Radioactivity, 53*, 5–8.

Polikarpov, G. G., Egorov V. N. (1981). Capability of marine ecosystems for removal of radioactive and chemical pollutants from the euphotic layer, *Bulletin of the Academy of Ukrainian SSR, 2*, 73–82 (in Ukrainian).

Polikarpov, G. G., Egorov, V. N. (1986). *Marine Dynamic Radiochemoecology*. Energoatomizdat, Moscow (in Russian).

Polikarpov, G. G., Kulebakina, L. G., Timoshchuk, V. I. & Stokozov, N. A. (1991). ^{90}Sr and ^{137}Cs in surface waters of the Dnieper River, the Black Sea and the Aegean Sea in 1987 and 1988, *Journal of Environmental Radioactivity, 13*, 25–28.

Polikarpov, G. G., Livingston, H. D., Kulebakina, L. G., Buesseler, K. O., Stokozov, N. A. & Casso, S. A. (1992). Inflow of Chernobyl ^{90}Sr to the Black Sea from the Dnieper River. *Estuarine, Coastal and Shelf Science, 34*, 315–320.

Polikarpov, G. G. (1998). Conceptual model of responses of organisms, populations and ecosystems to all possible dose rates of ionizing radiation in the environment, *Radiation Protection Dosimetry*, 75, 181–185.

Polikarpov, G. G., Mirzoeva, N. Yu., Kulebakina, L. G. & Egorov, V. N. (2008). Radionuclides ^{90}Sr and ^{137}Cs in the Black Sea aquatic organisms after the Chernobyl NPP accident. In G. G. Polikarpov, & V. N. Egorov

(Eds.), *Radioecological response of the Black Sea to the Chernobyl accident* (pp. 275–295). Sevastopol, Ukraine: ECOSEA–Hydrophysics (in Russian).

Robison, W. L., Conrado, C. L., Bogen, K. T., Stoker, A. C. (2003). The effective and environmental half-life of ^{137}Cs at Coral Islands at the former US nuclear test site, *Journal of Environmental Radioactivity, 69(3)*, 207–223.

Spellman, F. R. (1996). *Stream ecology and self-purification*, Technomic Publishing Co., Lancaster, Basel.

Takeda, H., Miyamoto, K., Fuma, S. (1998). Comparative evaluation of ecological effects of γ-radiation and UV-C-radiation using an aquatic microcosm. In: J. Inaba, Y. Nakamura (Eds.), *Comparative Evaluation of Environmental Toxicants – Health Effects of Environmental Toxicants Derived from Advanced Technologies* (pp. 121–125) Kodansha Scientific, Ltd, Tokyo.

Thébault, H., Rodriguez y Baena, A. M., Andral, B., Barisic, D., Benedicto Albaladejo, J., Bologa, A., Boudjenoun, R., Delfanti, R., Egorov, V., El Khoukhi, T., Florou, H., Kniewald, G., Noureddine, A., Patrascu, V., Pham, M. K., Scarpato, A., Stokozov, N., Topcuoglu, S. & Warnau, M. (2008). ^{137}Cs baseline levels in the Mediterranean and Black Sea: a cross–basin survey of the CIESM Mediterranean Mussel Watch Programme. *Marine Pollution Bulletin, 57* (6–12), 801–806.

Thomas, A. J. (1997). Input of artificial radionuclides to the Gulf of Lions and tracing the Rhone influence in marine surface sediments, *Deep-Sea Research II. 44*, 577–595.

Tsyban, A. V., Izrael, Yu. A. (1988/1989). Monoc programme: Scientific substantiation of the integrated global ocean ecological monitoring system, *Environmental Monitoring and Assessment, 11(3)*, 279–292.

Tsuchiya, K., Harashima, S. (1965). Lead exposure and the derivation of maximum allowable concentrations and threshold limit values, *British Journal of Industrial Medicine, 22(3)*, 181–186.

UNEP. (2002). *Global Mercury Assessment, overview of existing and future national actions, including legislation, relevant to mercury*, UNEP, New York.

Vernadsky, V.I. (1965). The chemical structure of Biosphere of the Earth and its environment. Nauka, Moscow (in Russian).

Zaitsev, Y. P.,Mamaev, V. (1997). Marine Biological Diversity in the Black Sea. A study of change and decline, United Nations Publications, New York.

In: Seawater
Editor: Mathew R. White
ISBN: 978-1-62808-606-5

Chapter 4

MACRO-IMAGINEERING A "FIFTH COAST" USA: SALTON SEA MEGAPORT AND SEAWATER CANAL

Richard Brook Cathcart
Geographos, Burbank, California, US

ABSTRACT

When the Saint Lawrence Seaway macro-project was completed in 1959, it opened a "Fourth Coast" for the USA. (Its East, Gulf and West Coasts were provided by Nature.) Excavation of a Canal connecting the Gulf of California, through Mexico to near the city of Mexicali and beyond to the USA would permit seawater flooding of the below-sea level Salton Sea, a briny, contaminated and smelly endorheic lake that is fast disappearing after its unnatural creation by man-made flash flooding in 1905-07. This chapter will fathom some mysteries of the Salton Sea, including some speculative informed comments that might seem like science-fiction or science faux. However, the author takes the discussion tack that a practical rendition of the possible future geographical facts for this inhabited and cultivated desert region of southern California and northern Mexico, namely "Alternative Zero", is more real-world than Peter D. Ward's *The Flooded Earth: Our Future in a World Without Ice Caps* (2010). Seawater can be a beneficent liquid, sometimes tranquil, soft and tractable. And, sometimes, it causes peril for humans, as well as economic losses. Not insignificantly, the San Andreas fault—the most well-known plate tectonic boundary in the world—parallels the easternmost edge of the existing Salton Sea; the bi-national seawater Canal and Megaport constructors, influenced by Walt Disney's "Imagineering" urban planning concepts, must make certain macro-project design allowances recognizing that geophysical reality. For Mexico, international commercial trade and tourism are the only industries that allow a net flow of wealth from richer to poorer nations. "Alternative Zero" has the potential to alter the climate of the Southwest as well as northern Mexico for the better.

INTRODUCTION

The newest ship type, "Super-Panamax" vessels are too big (in draft, length, width) to fit through even the centennial upgraded post-2014 operational Panama Canal. As a result, this fact reinvigorates a macro-problem facing all international oceanic merchant shipping, namely the increasing importance of all the obvious usual sea route choke-points (Lewis, 1992); new international sea trade route constrictions are created too! "The vessels, which offer both higher speeds and economies of scale, can call only at a small number of very large ports called mega-ports" (Fox, 2001). A few USA megaports exist, since the late-20th century, because cost-cutting "Just-in-time" manufacturing has become paramount to America's factory-based production schedules—containerships are the most essential of all invented vehicles to distribute parts and finished products sourced globally. Swiftly circulating steel shipping containers, collectively known famously as "the Box", are the chief 21st century means for humans to package various and sundry items in order to move them from a place of manufacture to a place of sale-consumption as cheaply as possible.

The Port of Los Angeles/Long Beach became classed as megaports after the Panama Canal was opened to commercial traffic early in the 20th Century. These adjacent seaports are man-made, unlike the distantly separated naturally magnificent San Diego and San Francisco harbors! In 1962, the combination cargo-passenger N. S. Savannah visited Los Angeles/Long Beach and San Francisco's globally competitive harbors; nuclear-powered ships have less aerial emissions effects on global climate than do ships burning fossil fuels (Fuglestvedt et al., 2009; Paxian et al., 2010). By 2012, the Mediterranean Shipping Company (founded 1970) sailed the first megaship, the MSC Fabiola carrying more than 12,000 containers, to Long Beach Harbor during March 2012. During the 21st Century, all types of seagoing ships are bound to become impressively bigger—indeed, some vessels are now so massive, such as the disastrously grounded enormous tourist-carting cruise-liner *Costa Concordia*, that some may register seismic signatures when and as they strike reefs or coastal land (Mucciarelli, 2012)!

For megaports, vast areas of warehousing are not required and, thusly, megaports also need not be present at crowded, highly urbanized and developed waterfront cities (Fan et al., 2010; Imai et al., 2013). Possibly 95% of the cargo volume of worldwide commercial exchanges happens through the agency of vessels plying our world's ocean. Geographos postulates the 21st century development of a gigantic linear seaway, more than 25 meters in depth and with a minimum 200 meter width, connecting the Gulf of California to a greatly enlarged volume, strictly of seawater, expanded Salton Sea. In other words, the "Salton Sea" will become worthy of its old, yet descriptively inaccurate, toponym since currently it is but an ugly brine lake undergoing an almost chaotic period of compositional change of its awfully odorous ingredients! (Figure 1.) If international shipping ever comes to frequent the Gulf of California sea-lanes regularly, on the way to or from the enlarged Salton Sea, then Geographos believes a new commercial shipping choke-point will have been created that owes it origin entirely to massive anthropogenic earth moving! Earth's seven billion people, extrapolated to the present-day USA Standard of Living, constitute an indicative guide to our species' potential for future coastal interventions. Final realization of the enlarged Salton Sea should be termed an anthropic inland coastal ocean-connected seawater body. In response to the eight suggested alternative ecosystem restoration programs of 2006, Geographos offers herein "Alternative Zero"—that is, an enlargement of the Salton Sea, bringing that briny and polluted body of water to sea level under the same toponym by deliberate flooding through a seawater Canal transiting Mexico and part of southern California. "Under all [8] alternatives, the Salton Sea will recede and some of the area that is currently under water will be exposed. Over time, these areas may contribute to increased dust emissions from disturbance and wind erosion" (Anon., 2006).

In a near-trivial sense, the navigable seawater channel building macro-project proposed is a larger version of the eight hectare seawater swimming pool built next to the Pacific Ocean by the Crystal Lagoons Corporation at the San Alfonso del Mar living complex in Chile! During the 1950s especially, the Salton Sea (unconnected to the Gulf of California) was a thriving California aquatic resort fostering swimming, recreational sailing and seasonally popular speedboat races (Anderson, 2011).

Powerboat enthusiasts often make the mathematically unverified claim that the summertime Salton Sea boasts the "fastest water in the world" because of low barometric air pressure and its high water density; apparently the speedboat propellers have a better "bite" in high-density water. The return of civilian water sports enthusiasts might be accompanied by new kinds of yachts—perhaps ship-like homes such as the 37 meter-long "Orsos Island", developed by Gabor Orsos? As of early-2013, super-yachts more than 100 meters in length are rare but, increasingly more common, with an estimated twenty-four on the water worldwide!

During 2012, the estimated annual evaporation from the Salton Sea's surface amounted to 1.8 meters of liquid, meaning that average speed of imported replacing seawater derived from the Gulf of California, by itself, flowing into enlarged pool through the artificial channel specified above as the Canal, would move northwards at about 0.09 meters per second. Predictable tidal and improbable tsunami effects on the Canal will require some costly computational modeling by experts. Of course, ships using the Canal would induce waves beating the sides of the channel that will be at least equivalent to those produced in the narrowest segments of the Saint Lawrence Seaway (Gharbi et al, 2010). It is just possible that a Canal electric mule system will be required, similar to that used—for a different purpose—at the Panama Canal where powerful mules [locomotives] guide rather than tow large ships which still use their onboard engines to move. In addition, naturally, litter retrieval and inland waterway maintenance must be done by roving marine debris-collecting floating and powered dustcarts.

Construction of the Canal—its official identifying name is still to be selected by those undertaking its realization with sufficient financing—would mean big-time earthwork excavation, 45 meters depth, extending from the Gulf of California about 113 kilometers to near Mexicali (founded 1903) and beyond through the USA's territory. For the 20th Century world-public, the Culebra Cut best represented the difficulty of digging Panama's original canal (Missal, 2008); the 21st Century's Salton Sea Canal will have no such awesome mountain impediment to serve the same psychological function! Some material could be lifted and shifted by nuclear-powered dredger (Cathcart, 2008), some by bucket-wheel excavator. The volume of earthy material moved to form a linear Canal cannot be less than 1,582,000,000 cubic meters of soil, rocks, silt, and anthropogenic stuff (operational facilities such as super-highways, roads, farms, desert towns, irrigation canals, natural gas pipelines, electricity transmission lines as well as civilization's castoffs, the humorously termed "California Conglomerate")! All removed earth will be used to form new, designed landscapes beside the Canal. It is possible, thereby, that a "linear city" urban form might spontaneously develop (Gauthier, 1995; Cathcart 1996); such a linear city could, later, expand into something appearing much like a macro-imagineered Disneyesque Amsterdam or Dubai, with canals as a major feature of their layout. Figure 1, below, shows two possible Gulf of California to enlarged Salton Sea routes for the Canal proposed herein.

Figure 1. Northernmost Gulf of California and the full delta of the Colorado River, lower right. The straight lines—not to scale in width—indicate the Canal's two most obvious routes serving the enlarged Salton Sea and its megaports. The channel on the left is the Baja California Channel which reaches the realized Salton Sea via the seawater-flooded Laguna Salada. It will require slightly more dredging than the Sonora Channel examined herein. The Baja California Channel would not ruin valuable farm and ranch lands, unlike the Sonora Channel. (Image: NASA, altered by RBC).

Construction workers for the Canal might be temporarily housed in air-conditioned shipping containers outfitted with four bunks and minimal accommodations; these semi-permanent pop-up architectures would be a bit cruder than the parasitic housing made available by Daniel Keith Ludwig (1897-1992), around 1967-68, when he built the paper pulp mill "Jari Project" in Brazil near the mouth of the Amazon River (van der Spaa, 1980). An avant-garde, temporary city formed entirely of shipping containers existed for two weeks at the Over het IJ theater festival, held annually in the NDSM shipyard in Amsterdam. [NDSM is the Dutch-language acronym for the "Dutch Dock and Shipbuilders Society" and the festival, celebrating its 20th year in 2013, adopts its name from that city's IJ waterway. Since the late-19th Century, the North Sea Canal has connected Amsterdam with Ijmuiden by allowing passage of ocean-going ships.]

The postulated nuclear-power dredger—since 1988 the Russian Federation operated a nuclear-powered, 260 meter-long, 9 meter draft Arctic Ocean-capable LASH [lighter (barge) aboard ship] carrier and containership, the recently retired MV Sevmorput—could also create "containment islands" (Coch, 1996) in the shallows offshore from the enlarged Salton Sea's shoreline, perhaps by using Patrick Michael Cashman's method outlined in his US Patent 4,103,502 awarded on 1 August 1978. These islands could be as imaginatively shaped as those built offshore of Dubai (Kumar, 2009; Cavalcante et al., 2011)! Built in 2009, the 223 meters-long, 41 meter beam, draught of 15 meters,Cristobal Colon is the most high-tech dredger operating anywhere during 2013, capable of excavating seafloor materials from a depth of 155 m. The largest bucket-wheel excavator, the Bagger 293, is capable of moving 240,000 cubic meters of material daily. The Bagger 293, assembled in 1995, is the largest mobile land vehicle ever built! The MV Sevmorput, removed on 31 July 2012 from the Russian Federation's Shipping Register, had a capacity to carry only 1,328 shipping containers, far too few to matter economically. The supposed nuclear-powered dredger might be converted to a seawater desalination plant after all necessary work it could do locally had been completed.

There are several comparable 21st century earthworks in progress: the 75 km-long Arabian Canal may soon be completed in the very much "macro-imagineered" modern-day Dubai. The Arabian Canal, if commenced and when finished, will be 150 meters wide and 6 meters deep, created by excavation of 1,100,000,000 cubic meters of soil and unconsolidated rocky material. The Arabian Canal is the largest and most complex work of canal macro-engineering in the Middle East since excavation of the Suez Canal (Finkl,

Pelinovsky and Cathcart, 2012). Its cost will be exceed US Dollars 10,000,000,000 to complete, with residential and office-space adjoining the Arabian Canal. Another canal macro-project, the planned 48 kilometer-long Canal Istanbul in Turkey, could be finished by 2023 to celebrate the 100th Anniversary of the Turkish Republic. The sea level Canal Istanbul will connect the Black Sea with the Sea of Marmara. The waterway, allowing ships to bypass the shipping-congested Bosporus route, will have a length of 45-50 kilometers, a navigable depth of 25 meters and a width of 150 meters. Therefore, its excavation will necessitate the mechanical removal, and deposition elsewhere, of more than 1,000,000,000 cubic meters of earth.

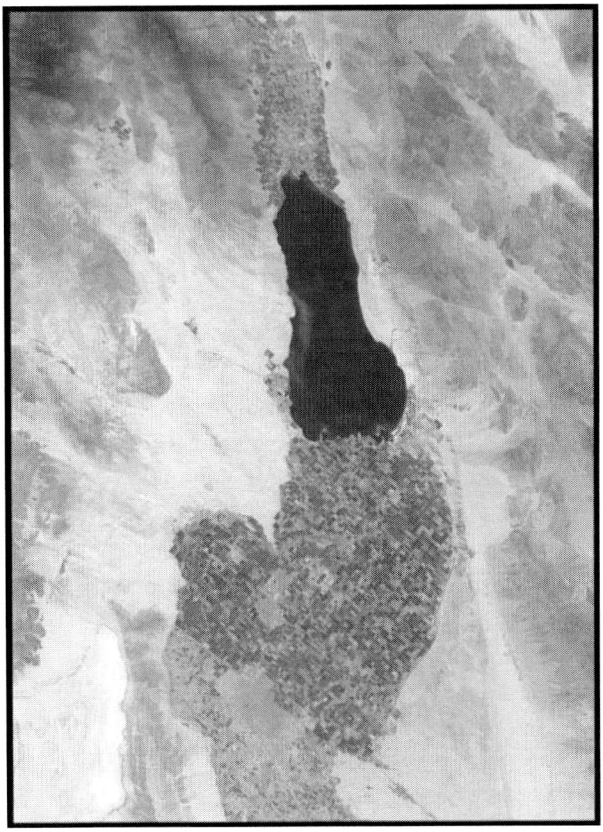

Figure 2. The Salton Sea as it more or less appears from an holistic outer space viewpoint today. With an area of almost 1,000 square kilometers, maximum depth of about 15 meters, and an elevation approximately 70 meters below sea level. (Image: NASA). In other words, The Salton Sea just about equals the area Japan has reclaimed from the ocean in the form of artificial offshore islands.

Figure 2, above, shows the existing Salton Sea. The enlarged Salton Sea's 21st century post-macroproject seawater volume, commensurate with "Alternative Zero", might be about 400 cubic kilometers, all seawater controllably let in from the Gulf of California to which the Salton Trough was many times linked during past geological time. (The extant Salton Sea should be filled to sealevel before the Canal is made to connect!) The Salton Sea's new, stabilized seawater area might become as much as 7,720 square kilometers, slightly more than seven times its present-day area (980 square kilometers), which equals the landscape surface area of the Salton Trough. (Figure 3.) Since previous incarnations of today's accidental Salton Sea have existed, at one time or another in the recorded and unrecorded past during the Earth's geological time,Geographos accepts the seawater filling of the Salton Trough as a restoration macro-project—that is, as "Alternative Zero". "Alternative Zero", though not nearly of the same geographical scale, might be likened to the early-1960s Ralph M. Parson Company-proposed macroproject for the arid northern Baja California region—namely, the North American Water and Power Alliance freshwater delivering "Baja Aqueduct" extending from Yuma to west coast freshwater reservoirs (Rolano, Guadalupe, Tijuana and El Compadre) in Mexico. "Alternative Zero" only slightly resembles "The Great Mexicali Energy and Shipping Canal" proposed by Roger Arnold (born 1944) of Silverthorn Engineering in Sunnyvale, California, during 2006 in his detailed blog-entry of 8th September 2010 (http://www. intelligentutility. com/article/10/09/great-mexicali-energy-and-shipping-canal). Both NAWAPA macroproject plans envision a huge increase in the volume of waste water Mexico sends yearly to the Salton Sea; already approximately 7.5% of all water entering the Salton Sea originates in Mexico and arrives via the New River's runoff; most of the New River's yearly runoff consists of untreated Mexico-generated urban sewage as well as pesticide and fertilizer-polluted agricultural runoff!

"The objective of the macro[-]engineering environmental restoration management process…is to increase the overall effectiveness by which organizational resources, committed to environmental restoration, are utilized", according to John Darabaris (2006). [As of 2012, Darabaris was working as Public Assistance Group Supervisor, Task Force Leader, at the Federal Emergency Management Agency (FEMA).] As Darabaris, others at FEMA and macro[-]engineers elsewhere know—especially after the risk mitigation and post-hurricane recovery organizational fiasco that was New Orleans before, during and after Katrina in 2005—the human ability and technological capacity to cause damage to our infrastructure and ourselves seems to have increased beyond our ability and capacity to forecast, ameliorate or reverse some forms of Nature's "wrath" pitted against humankind's infrastructure (Diacu, 2010).

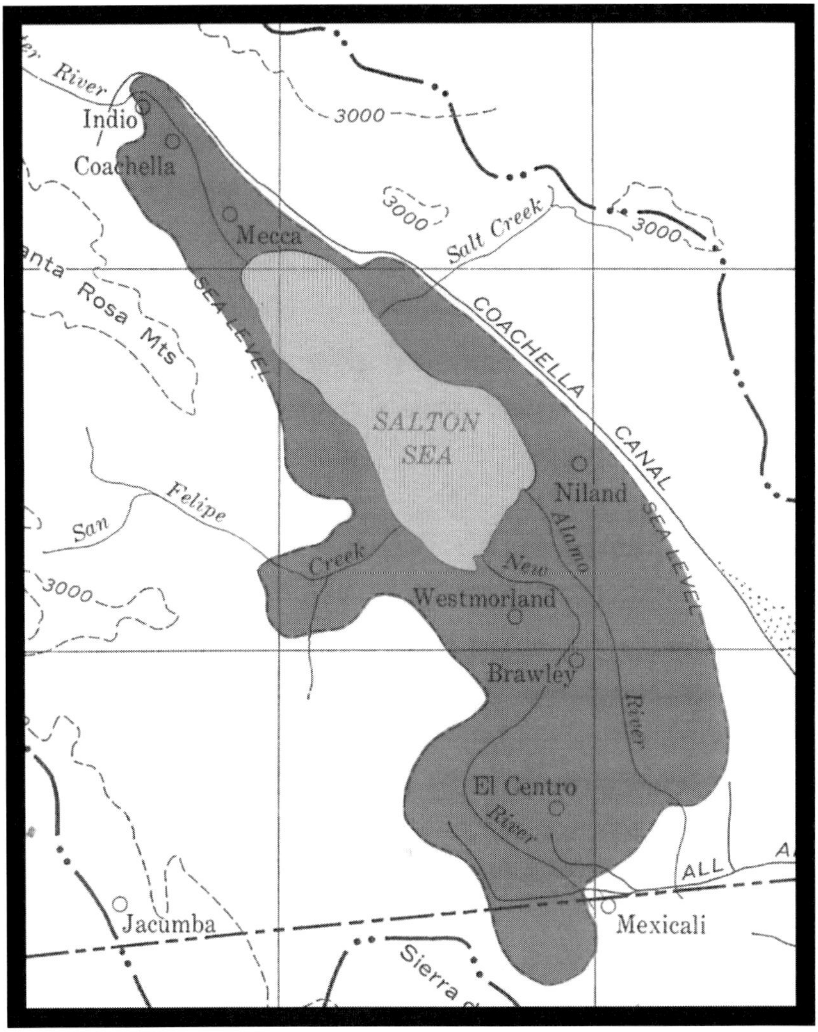

Figure 3. Light blue coloring represents 2013's Salton Sea. The dark blue color encircling, added to the surrounded light blue should just about represent the proposed enlarged Salton Sea. Contour line altitudes above sea level are expressed in feet. Seven especially noteworthy towns would be inundated permanently if the Salton Sea's water level were raised by about 70 meters. (Image adapted from various US Geological Survey maps by RBC).

Its open-water navigation depth would become a maximum 70+ meters; the bathymetric chart should be at least as good as the topographic maps existing before the enlarged Salton Sea's creation! Post-creation, a full tidal epoch (18.6 years) of measurements incorporating the full range of relative distances among the Earth, Moon and Sun must occur before truly accurate tide prediction tables could be issued! Before those measurements become fully available, a sophisticated array of finely crafted visible physical models—veritable working laboratory Lilliput's—as well as electronic supercomputer models will be used to study this evolving "Alternative Zero" Salton Sea semidiurnal natural event (Palmer, 2008). Moored ships do move when impacted by seawater waves, whatever their source (tide, wind or tsunami). Perhaps the megaport, sited somewhere in or on this new body of seawater, could be modeled on the Indian Ocean-facing Port of Salalah (16^0 56' 53" N. lat. by 54^0 0' 32" E. long.) in Oman, opened in 1998. Or, another option, modeled after Walt Disney's fabled and still immaterialized State of Florida city of tomorrow, EPCOT, only the post-EPCOT creation would be an aquatropolis, a city possibly named the Salton Sea Megaport centered on the new Canal (Gennaway, 2011; Sidaway, 2012; Brook, 2013)! Super-Panamax containership berths and other ship-servicing industries should be appropriately dimensioned and judiciously placed. As it did briefly during the 1950s, the Salton Sea could again support publicly popular fisheries like orangemouth corvina (*Cynoscionxanthulus*), croaker (*Bairdellaicistia*) and sargo (*Anistremusdavidsoni*).

Like Oman's Port Salalah, the Salton Sea Megaport would be located in a desert climate, which is well-suited strategically to help halt (I) illegal migration using uncomfortable/unsafe containers as temporary homes aboard vessels arriving from overseas and (II) prevent a port-detonated nuclear fission/fusion bomb from instantly causing damage to a surrounding highly-Blue Urbanized region or California city such as San Francisco and San Diego. In other words, millions of human lives in the State of California could be put less at terrorist-attack risk than should the Salton Sea Megaport not exist. [Interestingly, but for a seemingly minor incident of negotiated international treaty history, the Salton Sea Megaport would not today be needed because an All-American seaport featuring an oil-pipeline terminal, dry-bulk goods exporter, might have become a long time ago the western USA's ocean-shipping gateway to the markets of Latin America and Asia, sited on the delta of the Colorado River (Anon., 1983; Boime, Summer 2011)! Undoubtedly, there would have been a Mexico-USA disagreement over the acceptability of an international trade route within the Gulf of California (Szekely, 1977).]

Still, there are some noteworthy powerful natural geological hazards that might well be stimulated to new and deadly activity, if a volumetrically enlarged Salton Sea is ever accomplished. For example, some geoscientists allege that earthquakes in the badly fractured rock formations beneath the Salton Sea were triggered by influx of Colorado River floodwaters, as when ancient Lake Cahuilla existed. At its most recent maximum elevation, Lake Cahuilla rose to 12 meters above today's sea level between 100 BC and 1580 AD, covered 5,700 square kilometers with a maximum of 95 meters of freshwater depth (Buckles, Kashiwase and Drantz, 2002). The stress on the underlying rocks, riven by the San Andreas fault, by Nature's episodic lake loading, resulted in five major known powerful earth tremors measuring Magnitude 7 during a 1,000 year-long period (Brothers *et al.,* 2011). The report by Brothers and his colleagues (2011) unequivocally implicates a combination of Earth-crust loading and underground sediment layer fluid pore-pressure increases, complementarily associated with somewhat periodic Late Holocene freshwater flooding of southern California's Salton Trough by the Colorado River, in triggering generally horizontal (lateral) seismic fault shifts on the southern San Andreas. Those temblors occurred about 180 years apart. The most recent reported Magnitude 7 earthquake rattled Mexicali, Mexico, situated near the USA-Mexico International boundary on Easter Sunday in 2010. The future earthquake geo-hazard threat to the region may be lessened by that widely-experienced 2010 event, perhaps producing a long time period of seismic quiescence, but certainly the threat has not been removed forever! Macro-engineers are mindful that "The Gulf of California and the Salton Trough form the oblique-divergent boundary between the Pacific and North America [tectonic] plates from 22^0 to 34^0 N latitude" (Umhoefer, 2011). Also, ongoing anthropogenic changes to the region's geology, namely "Deep fluid extractions in the Cerro Prieto geothermal field...has caused [land] subsidence and induced slip on tectonic faults in the Mexicali Valley (Baja California, Mexico)" (Glowacka *et al.,* 2010). [Some geothermal fields (Mazzini et al., 2011) will be inundated by the Salton Sea's enlargement under "Alternative Zero" scenario: the Salton Sea, South Brawley, East Brawley and Heber fields, for example (Asmus, 2009); subsequently, these fields could be mothballed by capping or, possibly, operated as remotely-controlled underwater facilities. If used, however, geothermal energy might cause the prosperity of a large indigenous desalination industry.] Terrible earthquakes in the State of California are a known factor that must be taken seriously by public officials as well as the public (Robinson, 2012). On 18 October 2012 at 10:18 a. m. many concerned citizens participated in the "Great Southern California ShakeOut", a major earthquake preparedness drill for Californians. The epicenter for the 2012 imaginary "Big One" [of magnitude7+] was just north of the City of Thermal on the San Andreas Fault within the Imperial Valley. Figure 4, below is a digitized representation of the landscape situated between the northern part of the Gulf of California and southern part of the State of California in the vicinity of the Salton Sea, which is not imaged but would be located on the lower left-side of the image.

The impact of a gigantic artificial seawater emplacement, an actual new regulated sea-arm of the Gulf of California that is openly connected to the Pacific Ocean, by Canal digging passing near Mexicali, Mexico, leads to a subsequent situation of some interest to climatologists: what will be the effect of a true "Salton Sea" body of seawater on the desert of the State of California, State of Arizona, State of Nevada and the proud country of Mexico? In other words, the enlarged Salton Sea should be added to macro-engineering's list of climate macro-engineering proposals—intentional filling of Australia's Lake Eyre, Egypt's Qattara Depression, the seawater floodable chotts of Algeria-Tunisia and re-filling Central Asia's all-but-disappeared Aral Sea (Badescu and Cathcart, 2011)! It is proved that irrigation agriculture in the Central Valley has generated an anthropic modification of the aerial hydrologic cycle above the USA's Southwest landscapes (Lo and Famiglietti, 2013)!

The Colorado River's annual stream flow is expected to decline by 10% (Scanlon et al., 2012; Seager, 2012; Flint, 2012) and the San Diego County Water Authority announced during September 2012 a legal agreement to buy the entire output of what may become the New World's largest seawater desalination plant to be built at Carlsbad, California.

By 2020, the San Diego County Water Authority's macro-project planners want at least 7% of the reliable freshwater supply used in the City of San Diego and the immediate suburban region surrounding the city to be derived from the estimated one billion US Dollar seawater-desalting plant. The coastal Carlsbad plant, the Western Hemisphere's largest, should be delivering product by 2016, tentative construction costs of the plant and a pipeline inland to San Marco are estimated at one billion US Dollars. Of 225 USA metropolitan regions surveyed, in aggregate making the bulk of the Nation's population, San Diego was judged to be place-holder number 221, with a "high" water vulnerability rating (Padowski and Jawitz, 2012). Still ongoing are several bi-national, trans-boundary desalination macroproject study efforts:

(1) the Puerto Penasco Desalination Plant in Sonora transporting bulk potable product to Arizona via long-distance pipeline and (2) the Rosarito Beach Plant in northern Baja California intended to pump potable product cross-border to San Diego.

During 2012, the elected governments of the USA and Mexico signed an agreement of five years duration, Minute 319, an amendment to the 1944 treaty that gave Mexico rights to approximately 10% of the Colorado River's average annual flow, to share shortages and surpluses in the river's Lower Basin. This amicable agreement seems to be a step away from the usual hydro-hegemony negotiating tactics that has prevailed in the past. Ongoing climate change and human population increases in both countries, is forcing a costly 21st Century development of new reservoirs (mainly in the Colorado River's Upper Basin) and construction of desalination plants (mainly in the Lower Basin) as well as changes to long-term international operational compacts such as the 1944 treaty.

Freshwater demand by consumers in both nations is expected, by 2060, to be 25% more than the Colorado River can ever naturally provide, according to the 88-page December 2012 U.S. Department of the Interior, Bureau of Reclamation's Study Report "Reclamation Managing Water in the West: Colorado River Basin Water Supply and Demand Study"!

Figure 4. Topography of the landscape on which this essay is focused. The red-colored star indicates the epicenter of a magnitude 7. 2 earthquake that happened on 4 April 2010. The star is located approximately 64 kilometers south of the USA-Mexico boundary-line (Fregoso-Becerra and Garcia-Abdeslem, 2011). (Image: NASA).

DESERT LAKE MADE BY MACRO-ENGINEERING MISTAKE

During 2012, the Salton Sea was a shallow, endorheic, polymictic, saline lake in southern California, the largest lake in the State of California. Its hallmark salinity—now slightly more than open-ocean water—was first reached circa 1911. Before 1911, it was a freshwater lake created accidently by incorrect macro-engineering operation of a built region-wide irrigation system based exclusively on the tapped inflow from the Colorado River Delta, issuing from a river head-gate structure installed near Yuma, Arizona (deBuys and Myers, 1999; Redlands Institute 2002). A section of Salton Trough, over a long period of time, had been filled by accumulated Colorado River sediment, creating a transverse alluvial dam (the "Delta" proper) stretching east-west across the Salton Trough. The composition of the Salton Sea's water changed because, over time, there was widespread dissolution of previously-laid salt-beds when Lake Cahuilla and other, unnamed earlier lake incarnations existed during the Pleistocene, centuries before the Europeans arrived as explorers. (At its maximum volume, Lake Cahuilla was perched 12 meters above sea level and it may have required about 12 to 20 years of total present-day Colorado River flow to fill it to that specific elevation!)

Charles R. Rockwood formed the California Development Company in 1896 to direct some of the Colorado River's runoff into the Imperial Canal the company built by 1901. It was the collapse of the head-gate under flood conditions in June of 1905, and the subsequent unnatural diversion of the entire runoff of the Colorado River from 9 August 1905 into the Salton Trough, that altered the region—some of the Imperial Valley was lost to a gigantic downhill freshwater inundation to become the Salton Sea. The velocity of the flow was determined by gravity; on level ground the water's maximum flow was limited by the friction of the flat dry soil beneath, of the hot air above, and even within the water itself so that the anthropogenic flash flood could never move overland more than about 35 kilometers in an hour's time! The Colorado River's flow to the Gulf of California, away from the Salton Trough, was not fully restored until 11 February 1907. At least 12,000 people were made homeless virtually overnight as their properties and personal valuables were rapidly inundated by a flash flood-like runaway river (Sperry, 1975). It is possible that events like those of 1905-1907 could recur: if, for example, an "atmospheric river"-caused precipitation pelted the unfixed watershed of the Colorado River as it has in the past, then massive and uncontrollable flooding could result! Interestingly, Dr. Joseph Pomeroy Widney (1841-1938) as early as January 1873 proposed to deliberately divert the Colorado River into the Salton Basin Trough to re-create a prehistorically large freshwater lake and to ameliorate the region's hot and dry climate (Hellyer, 1949). In other words, he desired an artificial "Lake Cahuilla" unconnected to the Pacific Ocean; so, much-prized freshwater would be subject to intense evaporation in the hot desert!

Control of the runoff from the watershed of the Colorado River and its tributaries has been vastly improved, during the century or so since the early 20th Century, by almost innumerable dam emplacements, canal digging and levee construction macroprojects. These completed infrastructures underpin a unique, generally agricultural society situated in America's Southwest. An American sociologist has gone so far as to posit the credible thesis that today's highly controlled, technologized, Imperial Valley may be a kind of cyborg rather than a mere drainage basin or American cultural region (Rudy, 2005)! However, there is a serious, impending macro-problem facing the 400,000+ people living in the Salton Trough and their social and political leaders: climate change research appears to strongly indicate less total yearly freshwater runoff will flow downhill towards the Gulf of California during the 21st Century (Barnett and Pierce, 2009)! If future dry times do impact the 130 kilometer-long All-American Canal (completed in 1942)—the most capacious (740.6 cubic meters per second) irrigation canal in the world—will likely receive less freshwater to serve the needs of the Imperial Valley—indeed, it is the Imperial Valley's ***only*** significant freshwater source. Perhaps only clairvoyants can devise what the Salton Sea's future is because the whole enveloping social-geological situation, both by Nature and American/Mexican societies, is fantastically complex! Team Geographos has cogitated a bit on a possible solution, a solution outlined herein. Those living near the Salton Sea have recently endured a whiff of one possible, unpleasant future!

On 10 September 2012, an airborne rotten-egg stench, the likes of which had never been experienced by so many Californians before, spread from the Salton Sea northwards to Riverside County and the San Fernando Valley, as well as the cities of Palmdale and Lancaster in northern Los Angeles County (Anon., 24 September 2012). Reminiscent of London's "Great Stink" of 1858, the smell was strong enough to make some affected contemplate the revival of the miasmatic theory of human disease! The ambient pungent air, wafted by gale-like desert winds from off the Salton Sea, was caused by hydrogen sulfide gas, a waste product spewed by decaying organics brought to the Salton Sea's surface by wind set-up—the slight tilting of the entire briny water surface of the Salton Sea. [In some instances, say at the Black Sea, postulated concentrated hydrogen sulfide eruptive gas clouds could kill many people, possibly even during the 21st Century (Isvoranu and Badescu, 2011).] The fact that shifting earthquake-generating faults, associated with the San Andreas Fault Zone, are present and active also may add to the number of stimulations that caused the "Big Stink" of early-September 2012. Toxic fog afflicts people living in and near Salt Lake City, Utah almost every January—the U.S. Environmental Protection Agency has labeled the phenomenon the Nation's worst polluted air nearly every year. The periodic event worsens chronic asthma and other diseases but does not stink as strongly as the Salton Sea's so far sporadic bad air! "Alternative Zero" sea air will have gaseous and particulate air quality properties, including higher humidity, far different than currently persist above 2012's evaporating Salton Sea (Feistel et al., 2010). At least, such an effect cannot be totally discounted as irrelevant. Water quality has been of concern to those settled or working closest to the Salton Sea ever since its man-induced formation in 1905-1907.

Because the water body has no surface outflows, salinity in the Salton Sea is trending towards an increased concentration far beyond normal open-ocean, naturally variable "seawater". Added to the brine-water is agriculture's post-crop irrigation waste excess: selenium (Se), a sometimes dangerous-to-life trace element, and organic pesticides. Because the surface water layer becomes overheated during summertime it becomes stagnant and has very low values of dissolved oxygen, which can suffocate unfathomably huge schools of fish. Ghastly floating fish corpse mats and bobbing poisoned bird bodies are sporadically present on the ugly surface brine layer—although nothing like a seabird "madness" outbreak, as seen at Monterey Bay in 1961, has yet been observed at the Salton Sea! About the time that the Salton Sea was still considered attractive and healthful and a fun resort region, Hollywood's B-cinema producers entertained the public with *The Monster Who Challenged the World*—gigantic snails are "who"??—which starred the elderly, yet still virile, cowboy silent and talkies film star, Tim Holt. The bizarre, voracious human flesh-eating mollusks emerge from earthquake-opened cracks in the briny Salton Sea's bottom, and migrate into the freshwater-carrying All-American Canal, slowly expanding their "Snails Will Rule the World Again!" domain, ultimately to include the entire planet, of course, and endangering all humanity! Gadzooks! (Humans are victorious.)

Oddly, at the very time when the Salton Sea first became a widely-known and appreciated fishing, small boating and swimming resort in California's low-desert country, a future geological cataclysm was announced to magazine readers: a Los Angeles Department of Water and Power industrial training consultant during the period when Hoover Dam was being constructed to create Lake Mead on the Colorado River, Alfred M. Cooper penned and published during April 1950 an outlandish anthropogeomorphological prognostication for the Salton Sea's fate since that dam's completion (Cooper, 1950)—sudden enlargement of the Salton Sea by its *erosional* reconnection to the Gulf of California! [Warning: the publishers of Harper's Magazine, at http://www.harpers.org/archive/1950/04/0005584 appear to offer A. M. Cooper's article as a user-paid PDF download. However, while that PDF's pagination is correct, the magazine pages imaged as thumbnails do not match real-world, full-sized photocopier-preserved page versions actually made from the magazine's printed hardcopy.] Starting about 1950, humans invented the means (thermonuclear fusion explosives) to cause, or at least contribute, to the self-extinction of Homo sapiens and so Cooper's theme fitted the world-public's global atmospherics, it catered to its perceived influence in terms of his story's morals, mood, and attitude. Cooper's popularly-circulated statement was quickly refuted by geoscience experts as early as June of that same year (Lawson, 1950) who proved that the low-elevation silt barrier separating the Gulf of California and the "Imperial Valley"—a true real-estate promotion toponym, incidentally, first bestowed by the designer of the California Development Company's canal head-gate on the Colorado River, George Chaffey (1848-1932)—was not really being removed by the Colorado River's famous tidal bore.

The several massive concrete dams built to slow the downhill flow of the Colorado River (Hoover, Parker, Imperial, and Laguna) have, it is true, functioned as sediment traps—so much so that the delta of the Colorado River is no longer being maintained by sediment deposition (Huh, Coleman and Braud, 2007). So, erosion of the Colorado River's delta, like that of Egypt's Nile River, is ongoing, accentuated by a tidal bore associated with a twice daily change of sea level of about 10-15 meters. Team Geographos views this minor threat to valuable property in both Mexico and the State of California as one that can be alleviated by the construction of the Salton Sea Canal and the riprapping of the channel which carries a small volume of Colorado River outflow still. As Team Geographos perused the macro-problem, it came to our minds that a vast geographically enhanced Salton Sea, created by long artificial canal(s?) would make a near-perfect site for a nuclear-powered desalination complex of enormous output (Philbrick, 1975). In effect, the many macro-engineering plans that have been drawn-up over many decades to enhance the USA-Canada Great Lakes with freshwater imports from Canada's James Bay-Hudson Bay inspired this macro-engineering concept. If developed, a Salton Sea Desalination Complex could provide electrical power to much of America's Southwest and Mexico. Even the increasing freshwater needs of Las Vegas, Nevada might be satisfied without groundwater mining. Furthermore, the large inland sea created could add moisture to the overhead passing arid region air-masses by evaporation, thus potentially increasing downwind precipitation (rain and snow) as Dr. J. P. Widney hoped.

Connected to the Gulf of California by the Canal, the Salton Sea Megaport must be adaptable to future global sea level rise just like the East, Gulf, and West Coasts of the USA and Mexico (Tebaldi et al., 2012; Strauss et al., 2012). The Canal will likely present shipping interests with some slight landslide hazard macro-problems that incur not only blockage of a too-narrow Canal but also tsunami waves generated by landslides entering narrow bays and channels. Furthermore, a numerical and field study of ship-induced waves along the Canal should also be undertaken, using supercomputers and physical models. Ship-instigated waves nearly always cause bank erosion and can affect the stability of other nearby watercraft and unaware persons strolling or sitting on the shores! Perhaps, in some specific instances, the use of heavy-duty surface effect cargo-hauling vehicles ought to be warranted (Dindinger, 1976)? Hovercraft-type vehicles might prove less damaging to the desert's ecology than paved highways and ballasted railroad tracks. These machines could use the Canal. It would certainly be a high-energy means of shipping, quite a contrast to timely ocean shipping which is the most energy efficient mode of long-distance transport for large quantities of goods. Still, hovercraft vehicles can travel at high-speed, versus the normal running speed of a containership of between 20 and 25 knots in the open-ocean but far slower in a waterway. Considering that as the 21st Century unfolds transport technology will progress, possibly automobile, railway and shipping will become partially or fully robotized, leading to a situation where southern Californians and northern Mexicans endure the blessings and curses of a, more or less, "robot cargo cult" metaphor (Rudy, 2005)! Autonomous robots—shipping vehicles of many types (water-craft, land-craft and aerospace-craft—will have to be tightly controlled by human dispatchers.

The City of Bakersfield (35^{0} 22' N. lat. by 119^{0} 1' W. long.), at the southern end of the San Joaquin Valley, and the surrounding Kern County, has become an economic powerhouse of inland Central California because the location of the metropolis is within a three-hour automobile/truck drive of 90% of the State of California's 32,000,000+ residents! Bakersfield is a one-day turnaround for big-rig truckers delivering from San Francisco and San Diego. And, some petroleum industry experts suggest that an estimated twelve billion barrels of oil are to be found that can be withdrawn using shale fracking methods of mining, the largest deposit of any county in the USA! (By April 2013, the State Legislature was unwisely considering four bills in the Assembly banning fracking statewide, thus limiting inland Calfornians economic future.) The untapped by fracking Monterey shale field alone, which extends northeast more than 320 kilometers from Bakersfield, is estimated to hold more than 15.5 billion barrels of oil (Mills, 2012). Major new and renewed railway and highway routes, connecting Bakersfield with Barstow (34^{0} 53’ 36” N. lat. by 117^{0} 1’ 8” W. long.) might enhance these communities if those improved vehicle routes were, ultimately, terminated at the Salton Sea Megaport (Fan et al., 2010). Even pipelines, possibly used to export Kern County’s natural gas and oil via ships using the enlarged Salton Sea cannot be ruled out! So, too, are capsule-carrying freight pipelines and the pipeline transport of carbon dioxide as a gas (McCoy and Rubin, 2008). The employment of gaseous carbon dioxide as an effective deterrent to fish migration in channels is proven; a non-physical barrier of carbon dioxide gas could, therefore, be injected continuously into the incoming seawater flowing through the Salton Sea Canal in order to keep unwanted fish from entering the proposed “Alternative Zero” Salton Sea landscape configuration (Kates et al, 2012). All of these connective facilities will cross the long-proposed California-Nevada Interstate Maglev route. While no such magnetic levitation technology railroad, first tested in the 1970s, has yet hauled freight, in 2006 a 160 kilometer-long non-passenger route was studied by the Southern California Association of Governments to link the Ports of Los Angeles/Long Beach with inland cargo distribution centers (mostly consisting of numerous freight warehouses) at a cost then of 8.5 billion US Dollars.

When the much media-hyped San Andreas Fault eventually again rips the State of California, the Earth-crust’s shifting during the “Big One” will likely sever all East-West railway links. At the present time, about 97 kilometers of the Union Pacific track from San Bernardino to Yuma, Arizona runs roughly parallel to and, for about 47 kilometers, between 8 and 32 kilometers from, the San Andreas fault; the Union Pacific tracks proceed along the east side of the present-day Salton Sea, within 1.6 to 3.2 kilometers, until the lines cross the San Andreas fault. If there is a moving train in that place at the time of a major earthquake then there is surely a risk of catastrophic derailment. The Union Pacific track crosses the San Andreas Fault on a low embankment and the track there might be offset by as much as 5-6 meters! However, under Geographos’ “Alternative Zero” scenario, the artificially firmed ground carrying the Union Pacific tracks would be entirely submerged and all railway/San Andreas fault interaction would take place elsewhere during any

"Big One" seismic episode! Here, it ought to be noted that, at least in the popular electronic press, there is some e-news talk of establishing in the near-term future a "San Andreas Fault National Park" somewhere within the State of California (Manaugh, 2013); such sequestration, along with extant societal prohibitions (Lovich and Bainbridge, 1999), could constrain legally and politically any development of "Alternative Zero".

A Future Sealane and the Colorado River Delta

The Gulf of California (Lluch-Cota et al., 2007; Cudney-Bueno, 2009), possibly the only sea in the Earth-biosphere that is the national property of a single country, is an elongated basin of about 1,100 kilometers enclosed between the peninsula of Baja California and the mainland of Mexico except at its southern end where it is open to the Pacific Ocean. The Gulf's average depth is 730 meters. While the mainland seaport south of the Gulf's end, Lazaro Cardenas (17^0 55' 37" N. lat. by 102^0 10' 08" W. long.) was swamped by a 2. 5 meter-high tsunami on 19 September 1985, during 2012 it safely and efficiently handled approximately 1,242,777 shipping containers as well as other cargos. Historical information confirms that the generation of large tsunamis in the Gulf of California is improbable. Hurricanes originating in the Pacific Ocean rarely affect upper Baja California.

The Colorado River, as it meets the Gulf of California, forms a 2 to 8 kilometer wide, 50 kilometer-long estuarine basin, which opens at the distributaries into a 16 kilometer-wide delta. Numerous invisible submarine channels—virtually unmappable as they shift continually—and shoals project into the Gulf of California about 50 kilometers. The semidiurnal tidal regime has a range of more than 6.95 meters at the delta's mouths, sometimes as much as 10 meters. Such tides can cause seawater currents with velocities of 1.5 meters per second over the shallow, less than 30 meter depth, sedimentary sea-bottom adjacent to the delta, and more than 3 meters per second inside the estuarine basin. That is, the two main Colorado River outlets—the Baja California Channel and the Sonora Channel separated by Montague Island (31^0 44' 18" N. lat. by 114^0 44' 47" W. long.)—are tricky navigational challenges currently (Carriquiry et al., 2011); when tides approaching 10 meters do occur, 32 kilometer-long Montague Island is sometimes submerged! Therefore, controlled shipping lane approaches to the Canal, eventually leading northwards towards the Salton Sea, must become fixed and trained. Indeed, dredgers, whether nuclear-powered or otherwise must maintain an approach channel to the Colorado River's delta since the depths offshore are too limited (about 30 meters maximum) to allow the safe and uneventful passage of increasingly more massive containerships.

Two shipping entrances/exits are suggested: the first uses an improved Baja California Channel connecting to the rhombus-shaped Laguna Salada (32^0 33' 00" N. lat. by 115^0 22' 00" W. long.), a 60 kilometer-long and 17 kilometer-wide endorheic basin situated about 10 meters below present-day sea level situated in the hot desert (Dorsey and Martin-Barajas, 1999). Figure 5, below shows Laguna Salada as it might appear if filled to the extant global sea-level with imported seawater.

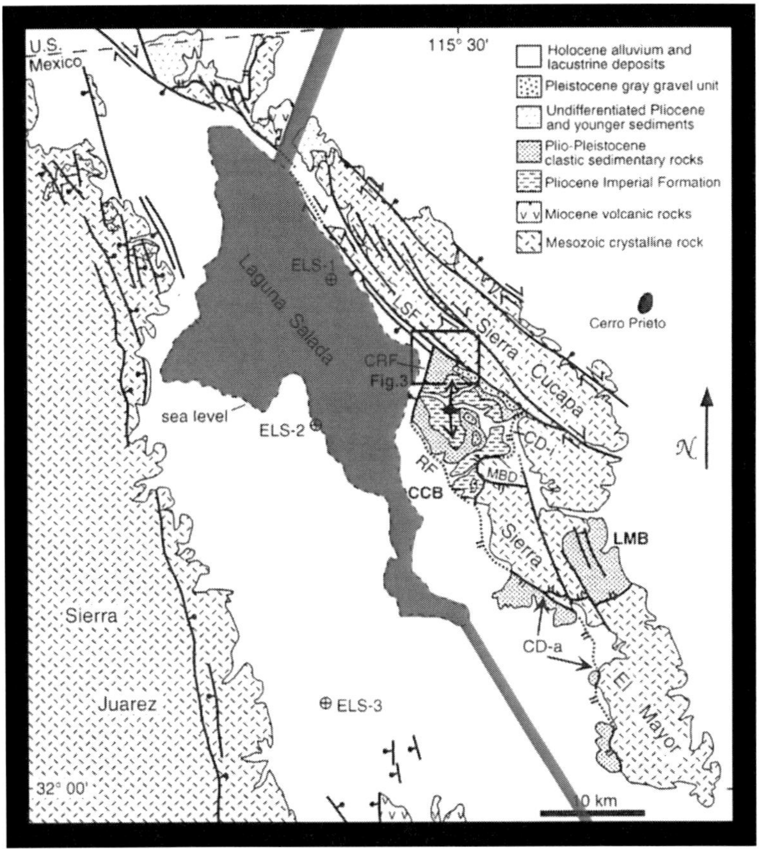

Figure 5. Filled by seawater arriving via the Baja California Channel (lower right part of the illustration), Laguna Salada could serve as a seaport for Mexico only or become a stop-over for bi-national shipping headed to or coming from the Salton Sea Megaport. (Image: US Geological Survey, altered by RBC).

Before the construction of Mexico's Highway Number 5 linking Mexicali and Tijuana, occasional high-tides in the Gulf of California inundated the Laguna Salada, making it shallowly wet. But, like the dry bed of the Salton Sea, the dry lake of Laguna Salada has endured winds of 170 kilometers per hour, blowing dust and imperiling human health. On 22 April 2012, the Discovery Channel filmed a sponsored television documentary that included video-tape of an unmanned, remotely-guided Boeing 727-212 crash test landing onto the dry Laguna Salada, west of the San Felipe-Mexicali Highway. The airplane wreckage was not salvaged (non-tragic litter). The second shipping entrance/exit, the improved Sonora Channel would, more or less, project in a straight line route generally northwards towards Mexicali, and the Salton Sea beyond. [This particular routing was selected in 2010 by Roger Arnold for "The Great Mexicali Energy and Shipping Canal" macroproject proposal.] These two channels offer each country some intriguing macro-Imagineering possibilities! For example, the Laguna Salada Megaport would serve Mexico exclusively or if only a Salton Sea Megaport is created, then it could become a bi-national facility! The proposed shipping channel linking the Gulf of California with Mexicali by way of the Laguna Salada could provide economical seawater-based heavy transport to and from the Gulf of California—and the Pacific Ocean beyond—from other important coastal cities such as San Diego in the USA, Ensenada in northern Baja California and even Phoenix, Arizona if the Salton Sea is never converted to an actual anthropogenic arm of the Pacific Ocean!

The Salton Sea at a Crossroads

Under what most Californians consider meteorological drought conditions, the Salton Sea's shrinking liquid water-covered area has exposed a former lake-bed wherein a century's aggregation of unpleasant and even sickening agricultural chemicals are pooled after running off the adjacent farm-fields and from return-flow irrigation canals. Even sewerage is bi-national! When completely dry, the highly comminuted solid lake-bed materials become airborne as dust clouds, lifted and carried by winds of merely 8 kilometers per hour! [10 September 2012's odiferous cloud named the "Big Stink" was pushed by North-blowing winds of almost 112 kilometers per hour!] Yet, in 2003, the Imperial Irrigation District, the agency which divvies the Colorado River freshwater brought to the Imperial Valley in the vital All-American Canal, reached a macro-problem resolution, titled "The

Quantification Settlement Agreement” (TQSA) that legalizes the transfer of a huge volume of Colorado River freshwater to the water-poor Southern California coastal cities of San Diego and Los Angeles. The so-called resolution allots no freshwater to maintain the Salton Sea’s level, making it physically possible for the Salton Sea’s actual disappearance over a short time period assured!

The Salton Sea’s absence in future leaves a contaminated lakebed exposed to pervasive aerial erosion, dust storms that could affect the health and welfare of the low-income people living near what was once the glorious Salton Sea. Even the wealthy, visitors as well as those living year-round in posh Palm Springs, will not be unaffected by toxic aerosols! Those persons with compromised cardiopulmonary systems will suffer almost endless assaults by vile dust clouds. Certainly, the entire multi-county region would be irrevocably impacted. The Imperial Irrigation District agreed in 2003 to deliver some undetermined volume of “mitigation water” to the Salton Sea until 2017, by which year that District supposedly will present to the State of California and the Federal Government a large-scale restoration macroproject plan that is readily implementable. “Implementation”, of course, is entirely dependent on monetary cost—that is, the cash-poor District could present a plan that is ridiculously costly and yet still fulfill its mandate. Impending is the real threat that the Imperial Irrigation District may terminate—by simple valve closure—all freshwater delivery on 1 January 2018. [Macro-project plants to construct energy farms and transmission lines through the desert of Southern California have all met with rejection, endless litigation and nil progress (Humes, 2009; Alagona and Smith, Fall 2012)—so far.] A State of California 3rd District Court of Appeals, on 8 December 2011, upheld the landmark Los Angeles-San Diego/Imperial Irrigation District TSQA. The legal wrangling continues, unresolved and, perhaps, unresolvable by ordinary restoration strategies! Despoliation and desert desiccrator-dessication can only result!

State of California Assembly member V. Manuel Perez, representing the increasingly dusty and smelly 856th District, during February 2011, sponsored legislation intended to transfer authority for the Salton Sea’s restoration, by whatever means necessary and taxpayer affordable, to the Salton Sea Authority. Perez’s bill in the Assembly died an ignominious death in the Appropriations Committee during August 2012. However, during January 2013 Perez introduced two Assembly Bills, AB 147 dealing with Salton Sea dust mitigation and AB 148 focused on renewable energy in the Salton Sea region. So, the way ahead legislatively in California is wide open, at least until the end of Perez’s term in 2014!

Conclusion

Michael J. Cohen and Karen H. Hyun authored *HAZARD: The Future of the Salton Sea With No Restoration Project* (Pacific Institute, Oakland, California, 2006). The most amazing section is titled "Why Try?" and simply poses the key question for all Salton Sea restoration advocates: "What if we don't try to restore the Salton Sea?" (GOTO: http://www. pacinst. org/reports/ saltonsea/index.htm.) Cohen and Hyun point out that by 2018 the surface elevation of the Salton Sea will become so low that approximately 68 square kilometers of lakebed will be exposed to the winds, lifting about 15.5 metric tonnes of sickening aerosols per day! They estimate that, by 2031, the weight of aerosols blown off newly exposed lake-bed (about 337 square kilometers) could generate 78 metric tonnes of breathable fugitive aerosols daily! This dreadful dusty future potential situation might be completely avoided if Team Geographos' macroproject plan to dig a Salton Sea Canal through Mexico, to just north of Mexicali, came to fruition! Without inundation by seawater, the evaporation and poisoned Salton Sea will become the single most important cause for the decline of real-estate values in the already financially distressed inland California region. All those involved in the care and use of the Salton Sea should make the greatest possible human effort to come to amicable agreements soon to inaugurate this landscape-saving macroproject concept! Geographos makes no claim to originating the concept, merely modernizing Dr. J. P. Widney's impractical idea of Salton Sea enlargement (with valuable freshwater) in a broader spatial context than ever previously disclosed by formulating the "Alternative Zero" all-seawater option. Stately submersion of some cities in the Salton Trough beneath a pristine, real Salton Sea composed of natural, wholesome seawater would, perhaps, erase many bad historical memories! In the process of flooding the Salton Trough with seawater, Americans and Mexicans will have created a fifth coastline as a boundary for North American.

References

Anderson, K. (2011) *The Salton Sea: Postcard History Series.* Arcadia Publishing: Charleston, South Carolina. 127 pages.

Alexander, L. M. (1992). The Role of Choke Points in the Ocean Context. *GeoJournal* 26: 503-509.

Alagona, P. S. and Smith, C. F. (Fall 2012). Mirage in the Making. *BOOM: A Journal of California* 2: 25-44.

Anon. (September 1983). Arizona's Lost Seaport. *Arizona Highways Magazine* 59: 47.

Anon. (May 2006). Salton Sea Update: Ecosystem Restoration Program. The Resource Agency, Department of Water Resources, Department of Fish and Game. Page 5.

Anon. (24 September 2012). Salton Stench. *Time* 180: 16.

Asmus, P. (2009). *Introduction to ENERGY in California*. University of California Press: Los Angeles. Map 7, page 140.

Badescu, V. and Cathcart, R. B. (Eds.) (2011). *Macro-engineering Seawater in Unique Environments: Arid Lowlands and Water Bodies Rehabilitation.* Springer: The Netherlands. 790 pages.

Barnett, T. P. and Pierce, D. W. (5 May 2009). Sustainable water deliveries from the Colorado River in a changing climate. *Proceedings of the National Academy of Science* 106: 7334-7338.

Boime, E. (Summer 2011). Navigating the Fluid Boundary: The Lower Colorado River Steamboat Era, 1851-1877. *Southern California Quarterly* 93: 175 et seq.

Brook, D. (2013). *A History of Future Cities*. W. W. Norton and Co : New York. 457 pages.

Brothers, D., Kilb, D., Luttrell, K., Driscoll, N. and Kent, G. (2011) Loading of the San Andreas fault by flood-induced rupture of faults beneath the Salton Sea. *Nature Geoscience* 4: 486-492.

Buckles, J. E., Kashiwase, K. and Krantz, T. (2002). Reconstruction of prehistoric Lake Cahuilla in the Salton Sea Basin using GIS and GPS. *Hydrobiologia* 473: 55-57.

Carriquiry, J. D., Villaescusa, J. A., Camacho-Ibar, V., Daessle, L. W. and Castro-Castro, P. G. (2011). The effects of damming on the materials flux in the Colorado River delta. *Environmental Earth Sciences* 62: 1407-1418.

Cathcart, R. B. (1996). Gauthier's 'Linear City'. *Environmental Conservation* 23: 286.

Cathcart, R. B. (2008). Kra Canal (Thailand) excavation by nuclear-powered dredges. *International Journal of Global Environmental Issues*, 8: 248-255.

Cavalcante, G. H., Kjerfve, B., Feary, D. A., Bauman, A. G. and Usselio, P. (2011). Water Currents and Water Budget in a Coastal Megastructure, Palm Jumeirah Lagoon, Ddubai, UAE. *Journal of Coastal Research* 27: 384-393.

Coch, C. A. (1996). Containment Islands: A Solution for Ports and the Environment. *Journal of Urban Technology* 3: 39-64.

Cooper, A. M. (April 1950). A Cataclysm Threatens California. *Harper's Magazine* 200: 66-69.

Cudney-Bueno, R., Bourillon, L., Saenz-Arroyo, A., Torre-Cosio, J., Turk-Boyer, P. and Shaw, W. W. (2009). Governance and effects of marine reserves in the Gulf of California, Mexico. *Ocean and Coastal Management* 52: 207-218.

Darabaris, J. (2006). *Macroengineering: An Environmental Restoration Management Process*. Taylor and Francis: Boca Raton, Florida. Page 3.

DeBuys, W. and Myers, J. (1999). *Salt Dreams: Land and Water in Low-Down California.* Albuquerque, NM: University of New Mexico Press. 307 pages.

Diacu, F. (2010). *Mega Disasters.* Princeton: Princeton University Press. 195 pages.

Dinginger, J. W. (April 1976). Strategic Mobility: Impetus for a Southwest Passage. *Proceedings: US Naval Institute*, pages 33-37.

Dorsey, R. and Martin-Barajas, A. (1999). Sediment and deformation in a Pliocene-Pleistocene transtensionalsupradetachment basin, Laguna Salada, north-west Mexico. *Basin Research* 11: 205-221.

Fan, L., Wilson, W. W. and Tolliver, D. (2010). Optimal network flows for containerized imports to the United States. *Transportation Research Part E: Logistics and Transportation Review*, 46, 735-749.

Feistel, R., Wright, D. G., Kretzchmar, H. J., Hagen, E., Herrmann, S. and Span, R. (2010). Thermodynamic properties of sea air. *Ocean Science* 6: 91-141.

Finkl, C. W., Pelinovsky, E. and Cathcart, R. B. (2012). A Review of Potential Tsunami Impacts to the Suez Canal. *Journal of Coastal Research* 28: 745-759.

Flint, L. E., Flint, A. L., Stolp, B. J. and Danskin, W. R. (2012). A basin-wide approach for assessing water resources in a semi-arid environment: San Diego region, California and Mexico. *Hydrology and Earth System Science*, 16, 3817-3833.

Fox, J. G. (2001) Sea Change in Shipping. *Proceedings US Naval Institute* 127: 62-65.

Fregoso-Becerra, E. and Garcia-Abdeslem, J. (2011). A possible seismomagnetic precursor of the El Mayor-Cucapah earthquake (Mw 7. 2) of April 4, 2010, Baja California, Mexico. *GeofisicaInternacional* 50: 211-225.

Fuglestvedt, J., Berntsen, T., Eyring, Vj.,Isakse, I., Lee, D. S. and Sausen, R. (2009). Shipping Emissions: From Cooling to Warming of Climate—and Reducing Impacts on Health. *Environmental Science and Technology* 43: 9057-9062.

Gauthier, G. (1995). The Linear City. *Environmental Conservation* 22: 273-275.

Gennaway, S. (2011). *WALT and the Promise of Progress City*. San Bernardino, CA: Ayefor Publishing. 372 Pages.

Gharbi, S., Valkov, G., Hamdi, S. and Nistor, I. (2010). Numerical and field study of ship-induced waves along the St. Lawrence Waterway, Canada. *Natural Hazards* 54: 605-621.

Glowacka, E., Sarychikhina, O., Suarez, F., Nava, F. A. and Mellors, R. (2010). Anthropogenic subsidence in the Mexicali Valley, Baja California, Mexico, and slip on the Saltillo fault. *Environmental Earth Science* 59: 1515-1524.

Hellyer, D. (June 1949). He Planned to Change the Desert Climate. *The Desert Magazine*, 12, 5-8.

Huh, O. K., Coleman, J. M. and Braud, D. (2007). The Colorado River Delta: Regional Geomorphology and Sedimentary Processes. *Journal of Coastal Research* 23: 1355-1358.

Humes, E. (2009). Solar Flare Ups: A fight over the future of clean energy is pitting environmentalists against one another. *California Lawyer* 29: 22-27, 5657.

Imai, A., Nishimura, E. and Papadimitriou, S. (2013). Marine container terminal configurations for efficient handling of mega-containerships. *Transportation Research Part E*, 49: 141-158.

Isvoranu, D. and Badescu, V. (2011) Peculiarities of tsunamis generated by asteroid impact or nuclear explosions in the northwest of Black Sea. *Natural Hazards* 58: 45-66.

Kates, D., Dennis, C., Noatch, M. R., and Suski, C. D. (2012). Responses of native and invasive fishes to carbon dioxide: potential for a nonphysical barrier to fish dispersal. *Canadian Journal of Fisheries and Aquatic Sciences* 69: 1748-1759.

Kumar, A. (2009). Reclaimed islands and new offshore townships in the Arabian Gulf: potential natural hazards. *Current Science* 96: 480-485.

Lawson, L. M. (June 1950). River and Tide Effects—Lower Colorado River. *Civil Engineering* 20: 31-33.

Lo, M-H and Famiglietti, J. S. (2013) Irrigation in California's Central Valley Strengthens the Southwestern U. S. Water Cycle. *Geophysical Research Letters*: DOI: 10. 1002/grl. 50108.

Lovich, J. E. and Bainbridge, D. (1999). Anthropogenic Degradation of the Southern California Desert Ecosystem and Prospects for Natural Recovery and Restoration. *Environmental Management* 24: 309-326.

Lluch-Cota, S. E., Aragon-Noriega, E. A., Arreguin-Sanchez, F. (2007). The Gulf of California: Review of ecosystem status and sustainability challenges. *Progress in Oceanography* 73: 1-26.

Manaugh, G. (18 January 2013). San Andreas: Architecture for the Fault. GOTO: "BLDG/BLOG" bldgblog. blogspot. com/2013/01/san-andreas-architecture-for-fault. html.

Mazzini, A., Svensen, H., Etiope, G., Onderdonk, N. and Banks, D. (2011) Fluid origin, gas fluxes and plumbing system in the sediment-hosted Salton Sea Geothermal System (California, USA). *Journal of Volcanology and Geothermal Research* 205: 67-83.

McCoy, S. T. and Rubin, E. S. (2008). An engineering-economic model of pipeline transport of CO_2 with application to carbon dioxide capture and storage. *International Journal of Greenhouse Gas Control* 2: 219-229.

Mills, M. P. (16 January 2012). California Could Be the Next Shale Boom State. *The Wall Street Journal* CCLXI: A13.

Missal, A. (2008). *Seaway to the Future: American Social Visions and the Construction of the Panama Canal*. Madison, Wisconsin: The University of Wisconsin Press. Page 69.

Mucciarelli, M. (2012). The Seismic Wake of the Costa Concordia. *Seismological Research Letters* 83: 636-638.

Palmer, A. C. (2008). *Dimensional Analysis and Intelligent Experimentation.* World Scientific Press: UK. 164 pages.

Paxian, A., Eyring, V., Beer, W., Sausen, R. and Wright, C. (2010). Present-Day and Future Global Bottom-Up Ship Emission Inventories Including Polar Routes. *Environmental Science and Technology* 44: 1333-1339.

Philbrick, A. K. (1975). Perceptions and Technologies of Determinants of Predictions about Earth, 2050. Chapter 2, Figure 2. 2 at page 27 *IN* Abler, R., Janelle, D., Philbrick, A. and Sommer, J., *Human Geography in a Shrinking World*. Belmont, California: Duxberry Press. 307 pages.

Padowski, J. C. and Jawitz, J. W. (2012). Water availability and vulnerability of 225 large cities in the United States. *Water Resources Research*, 48: W12529.

Redlands Institute (2002). *Salton Sea Atlas*. Redlands, CA: ESRI Press. 127 pages.

Robinson, A. (2012). *Earthquake: Nature and Culture*. London: Reaktion Books. 208 pages.

Rudy, A. P. (2005). Imperial contradictions: is the Valley a watershed, region, or cyborg? *Journal of Rural Studies* 21: 19-38.

Seager, R., Ting, M., Li, C., Naik, N., Cook, B., Nakamura, J., and Liu, H. (2012). Projections of declining surface-water availability for the southwestern United States. *Nature Climate Change*. DOI: 10. 1038/nclimate1787.

Scanlon, B. R., Faunt, C. C., Longuevergne, L., Reedy, R. C., Alley, W. M., McGuire, V. L. and McMahon, P. B. (2012). Groundwater depletion and sustainability of irrigation in the US High Plains and Central Valley. *Proceedings of the National Academy of Sciences*, 109, 9320-9325.

Sidaway, J. D. (2012). Geographies of Development: New Maps, New Visions? *The Professional Geographer* 64: 49-62.

Sperry, R. L. (1975). When the Imperial Valley fought for its life. *The Journal of San Diego History* 21: 1-9.

Strauss, B. H., Ziemlinski, R., Weiss, J. L. and Overpeck, J. T. (2012). Tidally adjusted estimates of topographic vulnerability to sea level rise and flooding for the contiguous United States. *Environmental Research Letters*, 7, 014033.

Szekely, A. (1977). The Patrimonial Sea to the Rescue of the Gulf of California. *Natural Resources Journal* 17: 113-122.

Tibaldi, C., Strauss, B. H. and Zervas, C. E. (2012). Modelling sea level rise impacts on storm surges along US coasts. *Environmental Research Letters*, 7, 014032.

Van der Spaa, D. (1980). JARI: A Billion Dollar Gamble. *National Geographic* 157: 686-711.

In: Seawater
Editor: Mathew R. White
ISBN: 978-1-62808-606-5

Chapter 5

ASSESSING THE EFFECT OF NEARSHORE CURRENTS OVER THE SO_2 ABSORPTION IN CARIBBEAN SEAWATER

C. Riverol[1*] and T. Alvarez[2]
[1]Chemical Engineering Department, University of the West Indies, St. Augustine Campus, Trinidad, Republic of Trinidad and Tobago
[2]ECCO Research Group, JC Engineering, Av. Andres Bello, Caracas, Venezuela

ABSTRACT

An evaluation of the SO_2 absorption in Caribbean seawater is measured by the addition of sulphuric acid at different conditions. The Caribbean is characterized by two main seasons: dry and hurricane seasons such that the variations nearshore currents are present. The measurements were carried out in closed systems, using hypothetical effluents where the nearshore currents were simulated. The results indicated that the addition of sulphuric acid at high speed current reduces the buffer capacity quickly and its acidity can increase until 6.6.

Keywords: pH; Seawater; sulphuric acid; solubility; buffer capacity; pollution.

* Corresponding author: Email: Carmen.Riverol@sta.uwi.edu.

1. Introduction

Sulfur is commonly present in seawater and marine sediments where the main sources of sulphate are the human activity (Shipping) and volcanic activity. The oceans cannot simply be regarded as a homogeneous system; the pH control occurs through equilibrium, a change in the proportions of the major elements in seawater should lead to a change in pH [1]. Specifically, the model for SO_2 absorption in seawater is based on a set of reaction in equilibrium as follow [2]:

$$SO_2(g) \Leftrightarrow SO_2(aq) \tag{1}$$

$$SO_2(aq) + 2H_2O(l) \Leftrightarrow HSO_3^-(aq) + H_3O^+(aq) \tag{2}$$

$$HSO_3^-(aq) + H_2O(l) \Leftrightarrow SO_3^{2-}(aq) + H_3O^+(aq) \tag{3}$$

$$CSO_3^-(aq) + H_3O^+(aq) \Leftrightarrow CO_2(aq) + H_2O\,(l) \tag{4}$$

$$CO_2(aq) \Leftrightarrow CO_2(g) \tag{5}$$

More details about the model can be found in [2] and [3]. Knowledge of the equilibrium data involving the sulphur-dioxide and sweater system is important for study and absorption process. According to reaction in equilibrium, the most important parameter in terms of SO_2 absorption in seawater seems to be the pH although other factors can be considered as well. The buffer capacity of the Seawater has a limit, when this limit is not respect the Seawater becomes more acid, in consequence, the fish and algae population can decrease slowly. This type of study in very important in the Caribbean because the petrochemical activities are very intense in Venezuela, Brazil, Mexico and Trinidad and Tobago. A limited literature data are found on solubility of sulphur dioxide in Caribbean Seawater and nothing about the influence of external parameters in the speed of adsorption. Some articles [3], [4], [5], [6] and [7] include measurements of sulphur dioxide in Seawater and distillate water at different temperatures, salinity and the impact on the environment however some discrepancies are even found and lack real scenarios are used in the simulation and experiments.

"The financial crisis has revealed a vulnerable global society. Fortunately, the latest developments have shown signs of recovery thanks to deliberate and well-coordinate global political actions which have created new confidence among companies nad consumers. This political commitment can be seen as recognition that global production and trade create wealth for all of us- with shipping and associated maritime industries as the primary enablers" [8].

Also, this article performs the buffer capacity of seawater in the Caribbean and the time necessary to reach pH-equilibrium after acid addition using hypothetical effluents. The measurements involve the seasonal effect (hurricane or dry) of the nearshore currents, that effect was created using different speed currents. The typical pH values find in the Seawater oscillate between 8.1 and 9.1 [6]. However the results in this article indicated that the pH can drop until 6.6 at high concentration of acid and high speed of the currents.

2. Material and Methods

These measurements were carried out in a stainless steel jacketed tank (4 L capacity) open-air system where the intention was to simulate Bays, Harbors e.g. small areas close to the Coast. The salinity of the water oscillates between 3% and 4% and it was measured using a salinity refractometer. The Seawater's composition is depicted in the Table 1. The water samples were collected at the north-west in Venezuela.

Table 1. Average Sea and Fresh Water Compositions

Ion	Freshwater (% weight)	Seawater (% weight)
Bicarbonate	48.33	0.40
Calcium	12.50	1.17
Chloride	6.67	55.19
Magnesium	3.33	3.70
Potassium	1.67	1.14
Silica	11.67	0.01
Sodium	5.83	30.72
Sulphate	10.00	7.68

The tank attaches several measuring devices as flow meter, pressure gauge, digital thermometer, pH meter and SO_2 analyzer. The experiments start at steady state conditions over a period of 2 hours where the experiments were setup at different sulphuric acid flow with a constant pH equal to 4. The jacket of the tank has a diameter of 0.15 m and a digital thermometer locates in the centre of the tank and uses circulating tap water as cooling flow. The nearshore currents were produced using small fans at the bottom of the tank. A total of 5 fans where used for simulating the currents. Each experiment was run 10 times. The reproducibility was high. The solutions were filtering and giving values within +/- 0.02 pH units. The temperature was considering constant because the tropical seawater temperature uses to oscillate between 23°C until 25°C. Before starting the measurements, the tank was filled with deionized water for 3 hours, after that, the vessel was emptied, dried and filling again with seawater. An additional time was need to stabilizing the temperature again. When the constant temperature was reached, the sulphur dioxide was bubble through the seawater until saturation conditions. Each experiment was run several times.

3. Experimental Results and Discussion

Measurements with natural Caribbean seawater allowed comparing the absorption of sulphur at different the speed of water and discharge points (each 0.5 m). The results are depicted in the Tables 2 to 5, where experimental results allowed testing the absorption of SO_2 in several conditions. Basically, the Table 2 indicates that the addition of 50 mg/L of sulphuric acid caused a pH shift of 0.2 units only.

The experiments demand an extreme degree of temperature control and very long period of time for each experiment thus it would have been more satisfying to have measured pH at true equilibrium rather than at steady state.

The field of coastal engineering is numerically complex and involving complicated model equations for interpretation of results. Mathematical models of wave propagation attempt to simulate the various mechanisms. The wave propagation affects the speed current and sulphur absorption. Over years, researchers have developed many elementary models [9], [10], [11] and [12]. However, except in some cases, the researcher has not prior knowledge of which of the mechanisms is dominating for a particular situation [13]. The results show that the addition of sulphuric acid decreases the pH until 6.6 at high speed of the currents; in consequence, the buffer capacity of the Seawater is seriously affected. Also these results can be interpreted as the buffer capacity of the Seawater can be affected when big traffic of ships (tankers or cruise) is traveling in a small area in the Caribbean.

Table 2. Impact he sulphuric acid on the pH seawater. Nearshore current 0.0 m /s

Sulphuric acid (mg/L)	pH
550	7.2
500	7.2
450	7.4
400	7.4
350	7.6
300	7.6
250	7.8
200	7.8
150	8.1
100	8.1
50	8.3
0	8.3

Table 3. Impact he sulphuric acid on the pH seawater. Nearshore current 0.08m /s

Sulphuric acid (mg/L)	pH
550	7.2
500	7.2
450	7.4
400	7.4
350	7.6
300	7.6
250	7.8
200	7.8
150	8.1
100	8.1
50	8.4
0	8.4

Table 4. Impact he sulphuric acid on the pH seawater. Nearshore current 0.20 m/s

Sulphuric acid (mg/L)	pH
550	6.9
500	6.9
450	7.0
400	7.1
350	7.4
300	7.4
250	7.7
200	7.9
150	8.0
100	8.0
50	8.3
0	8.3

Table 5. Impact he sulphuric acid on the pH seawater. Nearshore current 0.40 m/s

Sulphuric acid (mg/L)	pH
550	6.6
500	6.7
450	6.9
400	7.0
350	7.1
300	7.1
250	7.3
200	7.5
150	7.9
100	8.1
50	8.1
0	8.2

The models based on the conservation of mass and momentum are most widely used for predicting the transformation of waves in shallow water.However using experimental data, we tried to simplify the calculation using correlations that can predict the pH. A tendency for lower pH at lower alkalinity is observed as expected due to the corresponding lower neutralizing capacity. To establish the behavior of the system the experimental data of pH, temperature and SO_2 partial pressure were correlated as follow:

$$pH = 6.213 \times Uc^{0.12} - \frac{1601}{T} + \frac{8.974}{ln(P_{So2})} \quad (6)$$

Where the T is in K and the pressure in Pa, Reaction I is the dissolution of gaseous SO_2 in water and Uc is the velocity of the current. In the limit of low solute concentrations the amount of gas dissolved can be expressed by Henry's law which for SO_2 can be expressed as

$$p_{SO_2} = \frac{[SO_2(aq)]}{k_H} \quad (7)$$

A graph example of the correlation is shown in the Figure 1.

The most important parameter in terms of SO_2 absorption in seawater seems to be the alkalinity[10] although other factors may play a vital role as well [11] and [12]. In the following it will be investigated to which extent the inherent alkalinity in seawater increases its SO_2 absorbing capacity compared to pure water. The total alkalinity can expressed as:

A_T=$[HCO_3^-]+2[CO_3^{2-}]+[SO_2^{2-}]+[OH^-]+[HSO_3^-]+[CSO_3^-]...-[H^+]$.. and the ellipses represents insignificant contributions either unidentified or in neglectable amounts.

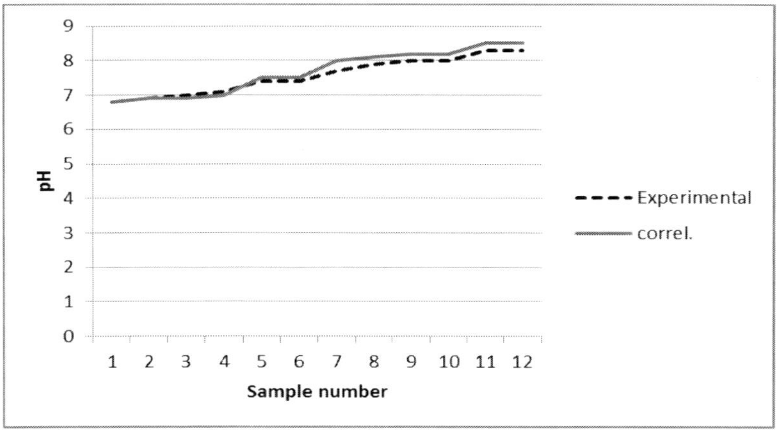

Figure 1. pH predicted vs. pH experimental.

Finally, in order to investigate the effect of variations in seawater salinity and alkalinity, a short correlation was developed using the above developed model. This correlation allows to predict the alkalinity in tropical water between 22-25 °C. For a given salinity the alkalinity is calculated using the equation (8) at constant seawater composition assumption.

$$A_T=165S^{1.1}+174 \quad (8)$$

Where S is the salinity.

CONCLUSION

The solubility of SO_2 in seawater has been studied experimentally at constant temperature in a range of speed of current between 0 m/s until 0.4 m/s. In particular, the influence of nearshore currents was studied subject to gas-exchange only. The results indicated that the impact of sulphuric acid can be dangerous when high concentrations at high speed current are discharged on the Sea reducing the buffer capacity and its acidity can increase until 6.6. The model could satisfactorily describe the dynamic changes in pH and the speed current of the Sea. It should be emphasized that the correlation includes no adjustable parameters that are determined so that the simulations match the experimental data.

In the other hand, this article tries to create consciences about sulphur pollution in small areas on the Coast (harbor, bay etc.) or any petroleum spill or explosions (human activity). In case of natural catastrophe, for example hurricane, we assume that the worst case never will be reached in the open ocean because the high speed currents will dilute the acid in the ocean sharply; is important to remember that the big petroleum companies shut down when the hurricane approaches category 4 or 5.

REFERENCES

[1] Wangerky, P. (1972). The control of seawater pH by ion pairing.; *Limnoly and oceanography*; January, vol. *XVII* (1), 1-4.

[2] Andersen, A. & Mayer, S. (2007). Use of Seawater Scrubbing for SO_2 Removal from Marine Engine Exhaust Gas. *Fuel and energy*, *21*, 3274-3279.

[3] Al-Enezi, G., Ettouney, H., El-Dessouky, H. & Fawzi, N. (2001). Solubility of Sulphur dioxide in Seawater; *Ind. Eng. Chem. Res.*, *40*, 1434-1441.

[4] Dacy, P. & Cope, D. (1986). Flue gas desulfurization system performance. IEA Research Report, London UK.

[5] Endressen, O., Bakke, J., Sorgand, E., Berglen, T. F. & Holmvang, P. (2005). Improved modelling of ship SO_2 emissions- a fuel based approach. *Atmospheric Environment*, *39* (20), 3621-3628.

[6] Capaldo, K., Corbett, J., Fischberk, P. & Pandis, S. (1999). Effects of ship emissions variations on sulphur cycling and radioactive climate forcing over the ocean. *Nature*, *40*, 743-746.

[7] Eyring, V., Köhler, H. W., Lauer, A. & Lemper, B. (2005). Emissions from international shipping: 2. Impact of future technologies on scenarios until 2050. *J. Geophys. Res*, 110-125.

[8] Shipping Green in the Future www.greenship.org, Denmark (2009).

[9] Bandyopadhyay, A. & Biswas, M. N. (2006). Prediction of the removal efficiency of a novel two-stage hybrid scrubber for flue gas desulfurizarion. *Chem Eng Technol*, *29*, 130–45.

[10] Amitava Bandyopadhyay & Biswas, Manindra N. (2007). Modeling of SO_2 scrubbing in spray towers, Science of the Total Environment, *383*, 25–40.

[11] Crusius, J. & Wanninkhof, R. (2003). Gas transfer velocities Measured at low wind speed over a lake. *Limnol. Oceanogr. 48*, 1010–1017.

[12] Gerbec, M., Stergarcek, A. & Kocjancic, R. (1995). Simulation model of Wet flue gas desulphurization plant. *Comput. Chem. Eng.*, *19*, s283.

[13] Wanninkhof, R. (1992). Relationship between wind speed and gas Exchange over the ocean. *J. Geophys. Res.*, *97*, 7373–7382.

INDEX

#

20th century, 128
21st century, 128, 129, 132, 134

A

accommodations, 132
acetic acid, 72
acid, x, 66, 72, 157, 158, 159, 160, 161, 162, 164
acidity, x, 70, 157, 164
AD, 137
adjustment, 104
adsorption, 102, 158
aerosols, 149, 150
aerospace, 144
aggregation, 148
agriculture, 138, 142
air quality, 142
air temperature, 6, 32
algae, ix, 94, 99, 118, 119, 123, 158
Algeria, 138
alkalinity, 162, 163, 164
Alternative Zero, x, 127, 129, 134, 136, 137, 142, 145, 150
ammonium, 70, 72
amplitude, 17, 31
ANOVA, 78, 81, 82, 83
anthropogenic influence, x, 94, 97
Appropriations Committee, 149
aquatic ecosystems, ix, 93, 94, 96, 118, 120
aquatic life, 100
arsenic, 66, 67, 78
ascorbic acid, 66
Asia, 136
aspiration, 60, 63, 66, 71
assessment, ix, 93, 95, 96, 101, 106, 118, 120, 122
assimilation, 100, 117, 120, 123
asthma, 142
asymmetry, 11
asymptotics, 46
atmosphere, 6, 7, 27
authority, 149

B

barium, ix, 54, 55, 56, 59, 62, 76, 78, 83, 88
base, 81, 82
biodiversity, 95, 119
biogenic salts, vii, 2, 6, 31, 32, 33, 34, 35, 36, 43, 46
biomass, 99, 102
biosphere, 146
biotic, 101, 102, 104
Black Sea region, 121
Brazil, viii, 53, 54, 56, 59, 81, 82, 132, 158

C

cadmium, ix, 54, 81, 88
calibration, 59, 63, 68
canals, 130, 148
capsule, 145
carbon, 145, 153, 154
carbon dioxide, 145, 153, 154
Caribbean, v, vii, x, 157, 158, 159, 160, 161
cash, 149
Cauchy problem, 14
cellulose, 59, 70, 73, 86, 89
Central Asia, 138
challenges, 146, 154
chemical, 31, 55, 56, 72, 96, 98, 101, 102, 104, 105, 118, 119, 124, 125
chemical properties, 56
chemicals, 95, 148
Chernobyl accident, 111, 123, 125
Chile, 129
chlorophyll, 107
chromium, 55
circulation, 95
cities, 129, 141, 148, 149, 150, 154
citizens, 138
City, 138, 142, 144, 151, 153
civilization, 130
clean energy, 153
cleaning, 96, 97, 105
climate, x, 120, 128, 136, 138, 139, 141, 151, 165
climate change, 139, 141
closure, 13, 26, 149
CO2, 154
coastal region, 57, 86
cobalt, 81
coding, 107
coefficient of variation, 78
color, 135
commercial, x, 67, 128, 129
communities, 118, 145
compaction, 103
composition, vii, 55, 64, 99, 140, 159, 164
computation, 100, 111
computational modeling, 130
conceptual model, 118
conditioning, 101, 103
configuration, 145
Congress, 123
conservation, 162
constant rate, 105
constituents, 55, 56
construction, 138, 139, 141, 143, 148
consumers, 102, 139, 159
consumption, 128
containers, 128, 132, 136, 146
contaminated and smelly, x, 127
contamination, 82, 83, 98, 103, 105, 113, 116, 117, 120, 122
Continental, 122
cooling, 70, 160
copper, ix, 54, 81, 88, 124
correlation(s), ix, 31, 36, 39, 54, 61, 62, 68, 76, 78, 89, 162, 163, 164
correlation coefficient, 61, 62, 68
correlation function, 39
corrosion, 56
cost, 128, 133, 145, 149
Court of Appeals, 149
covering, 72
cracks, 142
Croatia, 98
crop, 142
CRP, 97
crust, 137, 145
CT, 7
cycles, 95
cycling, 165

D

danger, 96
data set, 78, 81, 82
decay, 22, 72, 73, 95, 99, 100, 102, 104, 105, 108, 110, 111, 112
decomposition, 110
decontamination, 119, 120
deformation, 152
degenerate, 36
degradation, 42, 95, 102, 104, 105, 110

Delta, 140, 146, 153
Denmark, 165
deposition, 95, 96, 98, 101, 102, 133, 143
depth, vii, 1, 12, 13, 18, 20, 26, 59, 89, 98, 99, 129, 130, 132, 133, 136, 137, 146
detection, 61, 62, 65, 78, 86, 88, 89, 90, 120
deviation, 45
diet, 102
diffusion, viii, 2, 31, 32, 37, 39, 40, 45
diffusivity, 4, 42
digestion, 70, 71
dimensionality, 14
direct evaluation, ix, 93
direct measure, 59
disaster, 105
discharges, ix, 54, 57, 83, 90, 96
discontinuity, 4, 10
diseases, 142
dispersion, 4, 9, 56, 82, 89
dissolved oxygen, 142
distribution, 5, 10, 14, 18, 20, 22, 107, 108, 121, 145
diversity, 99
DOI, 154, 155
dose-response relationship, 118
dosimetric equivalents, ix, 94, 118
draft, 128, 132
drainage, 141
draught, 132
drought, 148
dry and hurricane, x, 157
dust storms, 149

E

earthquakes, 137
Easter, 137
ecological processes, 105
ecology, 121, 124, 125, 144
economic activity, 57
economic losses, x, 127
economic power, 144
economies of scale, 128
ecosystem, 94, 96, 99, 103, 104, 106, 129, 154
ecosystem restoration, 129
ecotoxicological, ix, 94, 98, 118, 123
effluent, viii, 54, 56, 58, 59, 74, 76
effluents, viii, x, 54, 76, 88, 89, 96, 157, 159
egg, 141
Egypt, 138, 143
electricity, 130
e-mail, 1
emission, 54, 59, 95
employment, 145
energy, vii, 2, 3, 4, 5, 6, 7, 8, 9, 10, 12, 13, 14, 15, 17, 18, 20, 26, 27, 28, 32, 34, 35, 44, 45, 46, 134, 137, 144, 149, 165
energy density, 45
engineering, 132, 134, 138, 140, 143, 151, 154, 160
enlargement, 129, 137, 143, 150
environment, 95, 100, 118, 120, 124, 125, 152, 159
environmental aspects, 91
environmental control, ix, 94, 116
environmental effects, 90
environmental factors, 105
environmental impact, vii, 56, 83, 88, 89, 95
environmental policy, 94
Environmental Protection Agency, 142
environmental standards, 118
environmental stress, 122
environments, ix, 93, 95, 96, 98, 99, 101, 103, 105, 106, 120
equality, 7, 35, 45
equilibrium, 5, 6, 7, 8, 10, 11, 45, 56, 72, 73, 158, 159, 160
equipment, 56
erosion, 105, 129, 143, 144, 149
ethics, 124
EU, 97, 117
European Commission, 95
European Community, 122
evaporation, 130, 141, 144, 150
evidence, 31, 82
evolution, 10, 12, 13, 22, 25, 105, 106, 107, 117, 120, 122
explosives, 143
exporter, 136

exposure, 118, 122, 125
extinction, 143
extraction, viii, 54, 55, 56

F

farms, 130, 149
fathom, x, 127
Federal Emergency Management Agency, 134
Federal Government, 149
FEMA, 134
filters, 58, 59, 70, 72, 73
filtration, 70
financial, 96, 159
financial crisis, 159
fish, 119, 142, 145, 153, 158
fisheries, 136
fishing, 142
fission, 136
flooding, x, 127, 129, 137, 140, 150, 155
fluctuations, 5, 32, 35, 36, 39, 42
flue gas, 165
fluid, 3, 5, 137
fluid extract, 137
food, 101, 102
food chain, 101, 102
force, 10
formation, vii, viii, 1, 2, 5, 31, 32, 53, 54, 55, 56, 66, 142
formula, 17, 19, 21, 35, 40, 117
fossils, 105, 119
freshwater, 106, 134, 137, 138, 140, 141, 142, 143, 148, 149, 150
friction, 5, 140
fusion, 136, 143

G

geographical, x, 105, 127, 134
geology, 137
geometry, 72, 73
Germany, 77, 98
GIS, 151
governments, 139
GPS, 151
graph, 163
gravity, 3, 7, 31, 32, 140
groundwater, 144
growth, vii, 1, 17, 22, 27, 41, 72, 102, 104, 113, 115, 123
Gulf of Mexico, 55

H

habitat(s), 95, 96, 101
half-life, 56, 64, 100, 102, 104, 106, 108, 109, 110, 111, 116, 125
harbors, 128
hazardous substances, 96, 99
hazards, 137
health, 94, 122, 149
heat conductivity, 13
heat transfer, 7
hegemony, 139
height, 6, 32
hemisphere, 28
high density polyethylene, 59
Highlands, 1
highways, 130, 144
history, 94, 98, 136
Holocene, 137
homeostasis, x, 94, 97, 119, 120, 121
homes, 130, 136
housing, 132
human, 95, 96, 97, 117, 120, 134, 136, 139, 141, 142, 144, 148, 150, 158, 164
human activity, 158, 164
human health, 148
humidity, 142
hybrid, 165
hydrocarbons, 55
hydrogen, 141
hydrogen sulfide, 141
hypothesis, 14

I

ideal, 65
IEA, 165
image, 138
immobilization, 95, 102
imports, 143, 152
income, 149
industries, x, 96, 128, 136, 159
industry, 56, 57, 90, 137, 144
infrastructure, 134
ingredients, 129
integration, 116
integrity, vii, 1
interface, vii, 1
interference, 63, 66
internal wave breaking, vii, 1, 4, 5, 6, 10, 11, 12, 16, 26, 28, 31, 34, 35, 36, 41, 45, 46
internal waves, vii, 1, 2, 3, 4, 5, 7, 9, 17, 20, 22, 28, 31, 32, 34, 35, 38, 44, 45, 46
International Atomic Energy Agency, 97
international trade, 137
ionizing radiation, ix, 94, 98, 118, 123, 124
ions, 55, 137
iron, 76
irrigation, 130, 138, 140, 141, 142, 148, 155
islands, 132, 133, 153
isotope, 63, 64
isotopes, ix, 54, 56, 66, 71, 76, 78, 86, 88, 89, 98, 105
Israel, 98

J

Japan, 123, 133
justification, ix, 94

K

kill, 141
kinetics, 98, 101
knots, 144

L

landscape(s), 131, 134, 138, 139, 145, 150
Latin America, 136
lead, 2, 55, 56, 71, 82, 120, 158
legislation, viii, 54, 125, 149
lifetime, 45
light, 55, 135
linear function, 33, 42, 43
litigation, 149

M

magnitude, 4, 10, 13, 27, 28, 31, 36, 43, 45, 86, 139
MAI, ix, 93, 112
man, x, 98, 105, 117, 120, 127, 128, 142
management, 134
manganese, 81, 82
manufacturing, 128
marine, ix, 57, 93, 94, 95, 96, 97, 98, 100, 101, 102, 103, 104, 105, 117, 118, 119, 120, 122, 123, 124, 125, 130, 152, 158
marine environment, ix, 57, 93, 97, 98, 101, 102, 103, 117, 120, 122
masking, 118
mass, 10, 32, 64, 65, 66, 71, 73, 83, 104, 117, 162
mass spectrometry, 66
materials, 31, 132, 148, 151
matrix, 56, 59, 60, 63, 66, 67
matter, 58, 83, 87, 88, 101, 103, 110, 132
measurement(s), x, 4, 22, 38, 44, 45, 46, 59, 64, 66, 72, 73, 98, 105, 106, 108, 136, 157, 158, 159, 160
media, 145
Mediterranean, 97, 105, 108, 122, 125, 128
mercury, ix, 94, 98, 116, 117, 118, 123, 125
Mercury, 125
metabolism, 98
metals, 55, 56, 59, 79, 80, 88, 97
metaphor, 144
meter, 129, 130, 132, 146, 160
methodology, 63, 65, 70, 71

Mexico, x, 127, 129, 134, 136, 137, 138, 139, 143, 144, 146, 147, 148, 150, 152, 153, 158
Middle East, 132
migration, 98, 101, 106, 136, 145
mixing, vii, 1, 3, 6, 8, 13, 22, 31, 32, 33, 36, 45, 81, 82, 105, 106, 108, 109, 121
modelling, 165
models, vii, 1, 6, 95, 97, 98, 136, 144, 160, 162
moisture, 144
mollusks, 142
molybdenum, viii, 54, 65, 82, 83, 88
momentum, 4, 7, 162
Moon, 136
Moscow, 46, 47, 48, 49, 50, 124, 125
Mussel Watch Program, 125

N

National Academy of Sciences, 93, 155
NATO, 97
natural gas, 130, 145
natural gas pipeline, 130
natural hazards, 153
natural resources, 124
nearshore currents, vii, x, 157, 159, 160, 164
nebulizer, viii, 54, 63, 64, 65, 67
negotiating, 139
Netherlands, 151
nickel, ix, 54, 55, 56, 76, 82, 83, 88
Nile, 143
non-uniformities, viii, 2, 28, 31, 32, 39, 40, 42, 46
normal distribution, 83
North America, 134, 137, 150
nuclear weapons, 105
nutrients, 55, 94
nutrition, 6, 31, 46

O

ocean upper layer spread, viii, 2
oceans, 101, 105, 122, 123, 158
OH, 163
oil, vii, viii, 53, 54, 55, 56, 57, 58, 74, 76, 77, 81, 83, 88, 89, 90, 91, 136, 144
oil production, vii, 55, 56, 57, 76, 81, 83, 88, 91
oil spill, 83
one dimension, 14, 18
operations, viii, 53, 81, 90
opportunities, 105
organisms, ix, 94, 101, 102, 110, 118, 119, 124
overlap, 26

P

Pacific, 50, 81, 82, 129, 137, 138, 141, 145, 146, 148, 150
pairing, 164
Panama, 128, 130, 154
parallel, ix, 10, 94, 98, 108, 118, 145
pathways, 94, 122, 123
PCBs, 122
permit, x, 127
pesticide, 134
petroleum, 90, 144, 164
pH, 55, 68, 70, 72, 157, 158, 159, 160, 161, 162, 163, 164
Philadelphia, 124
photosynthesis, 32, 43, 99
physical structure, vii, 1
phytoplankton, 32, 119
pipeline, 136, 138, 145, 154
plankton, 105, 119
plants, 139, 149
plate tectonic boundary, x, 127
platform, vii, viii, 54, 59, 73, 78, 79, 80, 81, 82, 83, 84, 85, 86, 87, 89
Pliocene, 152
plutonium, 105, 108
policy, 118
political leaders, 141
pollutants, ix, 57, 93, 95, 96, 97, 98, 100, 101, 102, 103, 104, 106, 110, 111, 112, 117, 118, 119, 120, 124

pollution, ix, 94, 95, 96, 98, 103, 105, 112, 116, 118, 120, 121, 123, 157, 164
population, 78, 82, 122, 123, 138, 139, 158
precipitation, 72, 140, 144
prediction models, 120
preparedness, 138
pressure gauge, 160
prior knowledge, 160
probability, 44
probability distribution, 44
produced water, viii, 53, 54, 55, 56, 58, 59, 60, 61, 62, 65, 70, 71, 73, 75, 76, 77, 78, 81, 82, 86, 88, 89, 90
producers, 142
prognosis, 123
project, x, 127, 129, 133, 134, 138, 146, 148, 149
propagation, 3, 10, 160
proportionality, 14
prosperity, 137
public health, 96
public officials, 138
pulp, 132
pure water, 163
purification, vii, ix, 93, 95, 96, 97, 100, 101, 102, 103, 104, 105, 106, 108, 110, 111, 112, 113, 116, 117, 119, 120, 121, 125
pycnocline., vii, 1, 31, 32, 45

Q

quantification, 86, 96

R

radiation, ix, 32, 94, 98, 118, 122, 123, 125
Radiation, 72, 91, 122, 123, 124
radioactive contamination, 99, 105, 121, 122, 123
radioisotope, 72, 103
radionuclide(s), viii, ix, 54, 55, 56, 59, 71, 76, 86, 87, 88, 89, 93, 98, 99, 100, 102, 105, 106, 107, 110, 111, 112, 116, 117, 118, 119, 121, 122, 123, 125
radium, ix, 54, 56, 59, 71, 73, 76, 78, 86, 88, 89
radius, 18, 20
rate of change, 114
reactions, 103
reading, 71
reality, x, 128
reasoning, 10
recognition, 159
reconstruction, 122
recovery, 70, 96, 98, 134, 159
recreational, 129
redistribution, 100
regions of the world, 89
rejection, 149
relatively narrow, vii, 1
remote sensing, 2
rendition, x, 127
renewable energy, 149
reproduction, 101
requirements, 6, 12, 33
researchers, 160
reserves, 152
reservoir, viii, 53, 54, 56
resistance, ix, 94, 118, 119
resolution, 63, 66, 71, 148
resources, 134
response, x, 12, 20, 35, 94, 95, 125, 129
restoration, 32, 134, 149, 150
rights, 139
risk(s), 95, 134, 136, 145
risk assessment, 95
room temperature, 72
routes, 131, 145
rules, 83
runoff, 106, 134, 140, 141
Russia, 77, 98

S

saline water, 88
salinity, 46, 63, 73, 95, 140, 142, 158, 159, 164
salt concentration, viii, 53, 55

salts, vii, 2, 6, 31, 32, 33, 34, 35, 36, 43, 46, 55
sample design, 59
saturation, 55, 160
scandium, 63
scattering, 4
school, 142
science, x, 127
sea level, x, 127, 129, 133, 135, 137, 140, 143, 144, 147, 155
sediment(s), 95, 96, 98, 99, 100, 101, 102, 103, 106, 110, 111, 122, 125, 137, 140, 143, 154, 158
sedimentation, 95, 96, 102, 103, 104
selenium, 66, 67, 142
self-purification, vii, ix, 93, 95, 96, 97, 100, 101, 102, 103, 104, 105, 106, 108, 110, 111, 112, 113, 116, 117, 119, 120, 125
self-similarity, 15, 18, 21, 46
sensitivity, 63
sewage, 134
shape, 9, 59, 118
shear, 3, 7, 10, 13, 22, 25
shoreline, 57, 132
shores, 144
showing, 57, 62, 67, 78, 81, 96, 100, 103, 106
signs, 159
silver, 55
simulation(s), 8, 159, 164
smoothing, 2, 4, 5, 28
smoothness, 31
society, 141, 159
sodium, 55
soil particles, 106
solid phase, viii, 54
solubility, 157, 158, 164
solution, 14, 15, 16, 18, 19, 21, 34, 35, 36, 63, 66, 87, 110, 113, 115, 141
sorption, 103, 105, 106
speciation, 124
species, 66, 95, 129
specific heat, 7, 27
speculative informed comments, x, 127
stability, 81, 144
stable states, 96
standard deviation, 45, 68, 83
state(s), x, 3, 6, 7, 8, 14, 22, 45, 94, 95, 96, 97, 99, 103, 104, 117, 119, 120, 122, 160
statistics, viii, 2
steel, 72, 128, 159
stock, 63, 66
storage, 154
stratification, vii, 1, 5, 6, 10, 12, 17, 22, 28
stress, 94, 113, 137
stress factors, 94
stressors, ix, 94, 118
stretching, 140
strontium, 56
structural changes, x, 94
structure, vii, 1, 16, 46, 96, 119, 125, 140
structure formation, 46
substitution, 36
sulfate, 72, 88
sulphur, 158, 160, 164, 165
sulphuric acid, x, 157, 160, 161, 162, 164
Sun, 136
suppression, 65
surface area, 134
surface layer, 106
sustainability, 154, 155
Sweden, 121
symmetry, 10, 15, 18, 20

T

tactics, 139
Task Force, 134
technologies, 165
technology, 144, 145
temperature, vii, 2, 6, 7, 8, 17, 26, 27, 28, 30, 31, 32, 33, 34, 35, 42, 43, 45, 160, 163, 164
territory, 130
testing, 105, 160
Thailand, 151
thermal expansion, 7, 32
thermodynamic equilibrium, 26
thermo-dynamical, vii, 1
thorium, 56

threats, 96
threshold level, 118
thumbnails, 143
tides, 146, 148
total product, 119
tourism, x, 128
toxic metals, 94
toxic substances, ix, 93, 95, 96, 98, 102, 103, 119
toxicity, ix, 93, 95, 96, 97, 112
trace elements, vii, viii, 54, 56
tracks, 144, 145
trade, x, 128, 159
training, 142
transformation, 95, 101, 102, 162
transmission, 130, 149
transparency, 95
transport, 46, 56, 94, 122, 144, 145, 148, 154
transportation, viii, 53
treatment, 73
Trinidad, 157, 158
Trinidad and Tobago, 157, 158
turbulence, vii, 1, 2, 3, 4, 5, 6, 7, 8, 9, 10, 11, 12, 13, 15, 16, 17, 18, 20, 22, 25, 26, 27, 28, 31, 32, 35, 36, 40, 45, 46
turbulent mixing, 26
turbulent patches, vii, 1
Turkey, 133
turnover, 94, 101, 103
twenty liters of seawater, viii, 54

U

U.S. Department of the Interior, 139
UK, 90, 121, 124, 154, 165
Ukraine, 93, 98, 125
UNDP, 121
uniform, 3, 6, 8, 10, 12, 17, 22, 26, 33, 44, 78
United, 125, 152, 154, 155
United Nations, 125
United States, 152, 154, 155
uranium, 64, 65, 83
urban, x, 128, 131, 134
USA, v, x, 1, 77, 127, 128, 129, 130, 136, 137, 138, 139, 143, 144, 145, 148, 154
USSR, 46, 47, 48, 49, 50, 51, 98, 123
UV, 125

V

validation, 70
valve, 149
vanadium, 56, 63, 76, 83, 88
variables, 13, 18, 20, 22
variations, x, 5, 22, 35, 157, 164, 165
vector, 37
vehicles, 128, 144
velocity, 4, 6, 7, 9, 10, 13, 17, 32, 140, 163
Venezuela, 157, 158, 159
vessels, 98, 128, 129, 136
vulnerability, 138, 154, 155

W

Washington, 90
waste, 55, 134, 141, 142
waste water, 55, 134
wastewater, viii, 53, 96
water quality, 96, 100
water quality standards, 100
water resources, 100, 152
watershed, 140, 141, 155
wave breaking, vii, 2, 4, 7, 10, 17, 28, 31, 35, 36, 46
wave number, 9, 38, 40, 41, 42
wave propagation, 10, 28, 160
wave vector, 5, 37
weak interaction, 3
weakness, 120
wealth, x, 128, 159
welfare, 149
well-being, 118
wells, 55, 56, 89
West Indies, 157
wind speed, 165
Wisconsin, 154
workers, 132

worldwide, 129, 130

Y

yield, 72

Z

zinc, 55, 56, 76, 124
zooplankton, 119